THE LIFE OF
·LTC·
ROLT

To David, Juliet and Rosy

THE LIFE OF
·LTC·
ROLT

WHERE ENGINEERING MET LITERATURE

VICTORIA OWENS

AN IMPRINT OF PEN & SWORD BOOKS LTD.
YORKSHIRE – PHILADELPHIA

First published in Great Britain in 2024 by
PEN AND SWORD HISTORY
An imprint of
Pen & Sword Books Ltd
Yorkshire – Philadelphia

Copyright © Victoria Owens, 2024

ISBN 978 1 39905 661 8

The right of Victoria Owens to be identified as Author of this work has been asserted by her in accordance with the Copyright, Designs and Patents Act 1988.

A CIP catalogue record for this book is available from the British Library.

All rights reserved. No part of this book may be reproduced or transmitted in any form or by any means, electronic or mechanical including photocopying, recording or by any information storage and retrieval system, without permission from the Publisher in writing.

Typeset in Times New Roman 10.5/12.5 by
SJmagic DESIGN SERVICES, India.
Printed and bound in the UK by CPI Group (UK) Ltd.

Pen & Sword Books Limited incorporates the imprints of After the Battle, Atlas, Archaeology, Aviation, Discovery, Family History, Fiction, History, Maritime, Military, Military Classics, Politics, Select, Transport, True Crime, Air World, Frontline Publishing, Leo Cooper, Remember When, Seaforth Publishing, The Praetorian Press, Wharncliffe Local History, Wharncliffe Transport, Wharncliffe True Crime and White Owl.

For a complete list of Pen & Sword titles please contact
PEN & SWORD BOOKS LIMITED
George House, Units 12 & 13, Beevor Street, Off Pontefract Road,
Barnsley, South Yorkshire, S71 1HN, England
E-mail: enquiries@pen-and-sword.co.uk
Website: www.pen-and-sword.co.uk

or

PEN AND SWORD BOOKS
1950 Lawrence Rd, Havertown, PA 19083, USA
E-mail: uspen-and-sword@casematepublishers.com
Website: www.penandswordbooks.com

Contents

Acknowledgements		vi
Preface		viii
Chapter One	Beginnings	1
Chapter Two	'the dirty devices of this world'	10
Chapter Three	The Apprentice in the Country	20
Chapter Four	The Apprentice in the Town	29
Chapter Five	Marking Time	40
Chapter Six	Love and War	50
Chapter Seven	Pacifism and the Path to Fame	61
Chapter Eight	Robert Aickman	74
Chapter Nine	After Lifford Bridge	87
Chapter Ten	The End of the *Cressy* Years	95
Chapter Eleven	The Talyllyn Railway	110
Chapter Twelve	Towards a Settled Life	122
Chapter Thirteen	Second Breakthrough	134
Chapter Fourteen	Chronicler of Companies	146
Chapter Fifteen	The Realm of Ghosts	157
Notes		168
Bibliography		202
Index		207

Acknowledgements

My especial thanks to Richard and Tim Rolt, Tom's sons and co-literary executors, for allowing me use of their father's letters, papers and published works. I am very glad to have the opportunity to thank them both for their help in the course of a happy afternoon spent in selecting pictures from the Rolt family's extensive collection of photograph albums for inclusion in this book's illustrations. Besides showing me over The Cottage at Stanley Pontlarge, his childhood home, Tim has been a great source of encouragement to me while I have been writing this book, always ready to answer my questions with unfailing goodwill. Tim's son, Felix Rolt, has generously shared his extensive knowledge of Rolt genealogy and family history, for which I am most appreciative.

I should like also to set on record my appreciation to Paula Mackersey for allowing me to quote from her late father's wonderful book, *Tom Rolt and the Cressy Years* (1985); to Peter White for permission to cite his unforgettable blog piece, *L.T.C. Rolt and a teaching career*; to Greg Morse of the Rail Safety and Standards Board for leave to quote his insightful remarks about *Red for Danger's* pertinence in the present day; to Christopher Awdry for allowing me to quote his late father's account of his experiences as a volunteer guard on the Talyllyn Railway; and to Will Willans for sharing family recollections both of Tom Rolt and Kyrle Willans.

Great thanks also to Nick Booth, Ironbridge Gorge Museums Trust's Head of Collections, to Sarah Roberts, IGMT's archivist and to Chris Pickford and his amazing team of volunteers who have undertaken the herculean task of cataloguing the Trust's extensive Rolt Collection. Also to Vaughan Pomeroy, President of the Newcomen Society; to Hannah Dale, archivist of Cheltenham College; to Ian Ferguson and his colleagues at Vintage Sports Car Club's library in Chipping Norton; to the staff of the Talyllyn Railway and to Keith Theobald at the Narrow Gauge Museum, Tywyn; to Louise Bruton, archives manager at the National Waterways Museum and Archive, Ellesmere Port; to the staff of Reading University Library's Special Collections' Archive of British Printing and Publishing; and to the National Archives at Kew. I wish to set on record my

Acknowledgements

appreciation to Sarah-Beth Watkins, my editor at Pen and Sword, and to Sarah Hodder my copy-editor.

Warm gratitude to the kind folk with whom I have been in contact while writing this book, notably John Anning; Dr Bruce Bomford; Joseph Boughey; Julia Elton; Julian Hunt; Peter Roberts and Raymond Russell. Special appreciation to John and Heather Freeman for their hospitality and encouragement, and heartfelt thanks to my husband David without whom it would never have seen the light of day.

Preface

Nearing the end of his life, Lionel Thomas Caswall Rolt, often known as Tom, wrote his autobiography. Consisting of three volumes – *Landscape with Machines*; *Landscape with Canals;* and *Landscape with Figures* – that it should have inspired much affection is hardly surprising. It tells the tale of a happy, productive life-well-lived; a life spent racing vintage sports cars, roaming the waterway network, managing an independent railway in Wales, and in prolific writing. A countryman who relished the vigour of industry; an astute engineer who wrote ghost stories; a reserved figure who became outspoken in print, his internal conflicts lent fire to his writing. Works of assured artistry, the *Landscape* books lay bare all the contrariety of Tom's existence.

Achievements aside, it is rather extraordinary that Tom's account should be almost the only record of his life story that we have. Ian Mackersey, admittedly, wrote an engaging if brief account of the years when Tom was married to Angela and lived on the canals – *Tom Rolt and the Cressy Years* (M and M Baldwin, London, 1985) – but the field is hardly crowded. After all, he was an inspired campaigner who spearheaded a revival of interest in the inland waterways, laid the foundations for the development of a lively heritage railway sector and, as a champion of what we have come to know as 'industrial archaeology', had a great gift for fostering an affection for the nation's industrial and engineering heritage.

Perhaps it stemmed from the fact that in the course of his life, he witnessed the loss of so much of it. Born in the Edwardian era, over the course of his lifetime he would see massive changes in agriculture (the sphere in which he first acquired the rudiments of engineering craft); in the design and manufacture of motor cars; on the waterways where the number of leisure craft came to exceed that of working boats; in the built environment, where post-war 'renewal' would encompass demolition in 1962 of the Euston Arch and on the railways with the phasing out of steam. Native mettle and a stubborn streak equipped Tom to resist the current of the times by championing the cause of history. From his early account of travel over England's decaying canal system at the end of the 1930s entitled *Narrow Boat* (which following its 1944 publication became a surprise bestseller) to his exploration of the history of the South Devon ball

Preface

clay industry in *The Potters' Field* of 1974, his narratives of past engineering endeavours and engineers' lives won many readers to his side.

He wrote at a time when authorship could still – just about – offer a plausible means of earning a living. Books during the 1940s, 50s, 60s and 70s remained a staple of life for many people. Provincial newspapers accorded generous space to reviews of new publications and most towns of any size had at least one independent bookshop. Before the need to provide space to accommodate computer terminals, the provision of books for information, recreation and entertainment stood at the heart of public libraries' operation. That is not to say that making a living from writing was easy; in all likelihood, it never has been. Nevertheless Tom, who reckoned that he produced '34 major titles over a period of 25 years' and lived to see more than 40 of his books published during his life was ready, even eager, to meet the challenges of authorship head on.

Since 2024 will mark the 50th anniversary of his death, a fresh appraisal of his life, his writing and his achievements feels timely. His adoption of the term 'Landscape' in the title of his autobiography may not be accidental. Not only does it reflect his deep and heartfelt love of countryside, but his trilogy is also, by any reckoning, a work of great artistry. Like a landscape painter, Tom shades and shapes the raw stuff of his memories to highlight some episodes while consigning others to silence in the shadows. While this book probably does not spring any major surprises concerning his life story, it may reveal something of the conscious art which he brought to bear upon relating his recollections.

Note

When writing about ships and steam locomotives Tom follows the convention of his time by referring to them with feminine pronouns as 'she' or 'her'. Nowadays, this custom has fallen into disuse and we tend to refer to vessels and engines impersonally as 'it'. In this book I have chosen to follow Tom's practice, partly because it avoids any awkward disjunction between my own text and my quotations from his works and partly because I like the old tradition.

I also follow him in referring to the Talyllyn Railway's coastal terminus as 'Towyn' although nowadays the town is known as Tywyn.

Victoria Owens,
Bristol, June 2023

Chapter One

Beginnings

Tom Rolt set his heart on a career in engineering when he was 4 years old. Out walking with his nurse, he saw *Katie*, a compact 0-4-0 15-inch gauge locomotive at work on the Duke of Westminster's private railway across Eaton Park on the outskirts of Chester. It was, he says, love at first sight and it would set his life's course. Captivated, he 'became a railway enthusiast from that moment'.[1]

Reflecting on his youthful zeal for 'all things mechanical' in *Landscape with Machines*, Tom finds it rather bewildering. Admittedly his great-grandfather, Henry Garnett, had financed the early business career of James Nasmyth, pioneer of the steam hammer and – although he does not mention it in the *Landscape* books – for a time, his father had a hand in managing the Hoole Bank Brickworks, but there was no family tradition of engineering.[2] Vocation, after all, tends not to pass down through successive generations in accordance with the principles of Mendelian genetics. Besides, Garnett's association with Nasmyth appears to have been more in the role of angel investor anticipating profit than any hands-on collaboration in the actual invention. Yet even if earlier generations of Rolts evinced little interest in machinery, they had other attributes which Tom, whether or not he cared to admit it, came to share. He does not, for instance, mention either his family's powerful love of language or their tendency to think for themselves. Since both attributes would play a considerable part in determining Tom's destiny, his inheritance from Rolts of earlier generations is worth at least some brief survey.

His great-grandfather, Colonel Sir John Rolt (c.1783–1856) was a career soldier who served under Wellington in the Peninsular War. An independent-minded man, in the 1840s, he wrote a book entitled *On Moral Command*. Purporting to be a manual of man-management for army officers, it carries a current of humane wisdom which Tom – himself, a life-long pacifist – took to be 'somewhat rare' in the context of nineteenth-century militarism.[3] Colonel Rolt had enough imagination to see things from the troops' point of view. He deplored, for instance, the army's practice of sending a raw recruit straight into his battalion, to be 'buffeted about by the old soldiers, and […] driven from post to pillar […] without one friend to cheer and support his spirits under the great change of position which he has lately experienced'. It would, he suggests, be better by far for the new man to be allowed four to six weeks under the care of 'the most steady, well-conducted and *best-tempered* [Colonel Rolt's emphasis] sergeant in the regiment' so as to get used to army discipline before taking his place among the seasoned soldiers.[4] If his great-grandfather's readiness to put himself into another person's figurative shoes is something that Tom shared, so too was his down-to-earth view of what makes for 'good fortune' in life. Far from owing anything to accidents of birth or chance, Colonel Rolt believed that 'advancement' – success, in other words – demanded 'downright hard work [...] close application, devoted zeal, determined gallantry and perseverance in conquering difficulties'.[5] For a writer as much as an army officer it was, give or take the gallantry, sound advice.

Colonel Rolt's son, Thomas Francis Rolt (1820–1901), may have intended to follow his father's career, for he joined the Coldstream Guards as an ensign in 1848. But he retired from his august regiment after only five years' service and married a woman named Mary Foot.[6] Before long the couple had a daughter, Edith Elizabeth, but Mary's health was none too robust and she died in 1858 in Algiers 'after a lingering illness borne with Christian fortitude'.[7] Thomas Francis remarried a year later, taking as his second wife Annie Garnett, the daughter of Nasmyth's old ally. From that time on, he devoted much of his time to 'hunting and procreation' which were, according to Tom, his 'favourite pastimes'. Since no fewer than eleven of Thomas and Annie's children survived into adulthood, the observation may not have been altogether wide of the mark. Many of them began their lives in the Rectory at Stow on the Wold which Thomas leased year after year on account of its being in Heythrop Hunt country.[8]

Yet he was perhaps a subtler man and certainly possessed a broader range of interests than his irreverent grandson allows. January 1865, for instance, found him in Stow on the Wold giving a dramatised reading from Dickens' *A Christmas Carol* with what the local paper termed 'great effect'.[9] An able draughtsman, he filled the sketchbooks that served him as the 'equivalent of the family photograph album' with hunting scenes, family portraits, drawings made in the course of his and Annie's travels in continental Europe and North Africa and, strikingly, a watercolour depiction of Brunel's Saltash Bridge in its construction phase.[10] Forceful in his opinions, the subject of Irish Home Rule – a much debated subject of the 1880s – would move Thomas to write stormy letters to the newspapers. 'Will you', he harangues the readers of the *Cheltenham Chronicle*:

> 'enrol yourself under the red flag of the Socialist and the Anarchist, and join the ranks of Mr Gladstone's friends, the Parnellites, the Fenians, the Moonlighters, the Boycotter? Or will you enrol yourself under the Union Jack of old England, under the Queen's colours of loyalty, order, and respect for law?'[11]

Beginnings

Gladstone, 'the Grand Old Man' who had introduced a Home Rule Bill in 1886, was a particular butt for his indignation. 'Let the G.O.M count up his years', he barracks in the *Gloucestershire Echo*:

> 'and remember that he holds no specific, more than any other mortal, for arresting the chariot wheels of the Juggernaut of Time from making sad inroads into even the highest intellect, and let him retire gracefully into the soothing calm of old age, employing time in the study of Homer, varied by collecting old china, or acquiring the Art of Jam making, but leaving the affairs of State to younger and now necessarily clearer minds.'[12]

Masterly in its rudeness, the passage rings with Thomas's sheer enjoyment of his command of language. If inherited talent counts for anything, it would serve Tom well, not that he was given to insults on this scale or, indeed, even cared to acknowledge the legacy. When Thomas Francis Rolt died in 1901, his estate was valued at the small sum of £625 gross. 'The horses', Tom surmised, had 'consumed the lion's share of the family fortunes.'[13]

As for Tom's immediate family, his father Lionel Caswall Rolt (1866–1941) known as Lily (pronounced 'Lylie') was Thomas and Annie's fifth son. As a young man, he had – Tom relates – travelled out to Australia to work on a cattle ranch. A lasting memento of those days was his stock-whip which, as a small boy, Tom watched him whirl and crack 'with a noise like a gunshot'; not only was it amusement for his young son, but it also satisfied Lily to reassure himself that he 'still had the knack'.[14] He spent some time in India where he worked on an indigo plantation and, in 1898, by now in his early 30s, joined the gold rush and headed for Yukon territory. What might have been an enriching adventure turned into a disaster; not only did Lily fail to find any gold, but his companions 'betrayed' him, with the result that, abandoned and penniless, he had to apply all his wit and stamina to survive.[15] Having reached Herschel Island, a polar centre of the whaling industry, he resourcefully offered to work his passage 'back to civilisation' on a whaler.

'Civilisation' meant his parents' home. Thomas and Annie Rolt were now living outside Chester, in the village of Christleton where Annie's brother Lionel was the rector.[16] Lily arrived sometime in the late summer or early autumn of 1899, having had his fill of adventuring and ready to appreciate domesticity and Cheshire's social and sporting life. In November 1899, he presented a 'handsome silver cup' at the Chester Cycling Club's annual dinner and the following spring saw him elected churchwarden of Christleton Parish Church.[17] Not only did it bring him to the forefront of parish life, with its parochial entertainments, but it also led to his becoming acquainted with many of the local clergy and their families.

Among them was Reverend Isaac Robinson Timperley, vicar of Guilden Sutton, his wife Mary, his stepdaughters and his young son.[18] They were what, at the present day, is known as a 'blended family'. Mary had been married before, her previous husband being one Thomas Clarke of Cogshall Hall, a wealthy gentleman significantly older than herself, by whom she had two daughters, Jemima and her sister Augusta.[19] Clarke appears to have died early in 1886, when the girls were no more than infants.[20] The following year, Mary married Timperley, whom the newspaper notice of their wedding describes as being 'of Slough, Bucks.'[21] Sometime later, they had a son – Herbert Luard.[22] Setting out on a career in the church, Timperley, like many other clergy, moved his wife and children from parish to parish. For a time, he held a curacy in Bracebridge near Lincoln; in 1891 he went as curate to Oulton in Suffolk; five years later in 1896, he took up the Guilden Sutton living and brought his family to Cheshire.[23] That they remained at Guilden Sutton for some ten years suggests that it suited them. It was perhaps in the course of attending North Cheshire's round of concerts, bazaars and flower shows that Jemima – later known as 'Mima' to her family and friends – became acquainted with Lily Rolt, sportsman, manager of the Hoole Brickworks, and warden of St James's Church, Christleton. In time they fell in love.

At much the same time, Mima's sister Augusta became engaged to a well-liked Chester GP named George Taylor and they married in February 1906. Their wedding took place in Wimbledon, since Timperley had recently taken up a post in the Anglican diocese of Southwark.[24] The Taylors – George's parents, his sisters and other relatives – were all present, but the *Cheshire Observer* which evidently sent a reporter to the ceremony remarked that it was 'of a very quiet character'.[25]

Drawing presumably upon memories of what he had heard from his parents, Tom claims that when Lily and Mima's courtship became a formal engagement, Timperley 'objected strenuously' to the nuptials. The fact that the wedding went ahead anyway (it took place at St Mary Without-the-Walls, Chester, on 24 September 1906) suggests that he did nothing so blatant as to declare the existence of a cause or just impediment, but his opposition made its mark, nevertheless. The Rolts visited George and Augusta Taylor at their Chester home every Christmas, and in summer stayed at Plas-y-Garth, their house in Denbighshire. Tom was fond of his aunt; he adored travelling in George's Daimlers and his cousins, Rosemary and Rosalind, were a friendly presence in his often solitary childhood. He also had touching affection for George's sister Hero, his godmother and his mother's lifelong friend whom he calls the 'kindest of women' and whose husband Kyrle Willans would be instrumental in securing his training as an engineer. But of Herbert, his half-uncle, he says nothing, beyond remarking that his mother's family 'did not figure largely' in his life when he was growing up.[26]

Beginnings

Behind this statement lurks a serious allegation. Somehow, Lily had 'discovered' that the Reverend Isaac Timperley – fretting, perhaps, over the injustice of his stepdaughters' benefitting from their father's munificence while no such bounty awaited his own son – had been 'cheerfully making away' with the monies that old Thomas Clarke apparently left to Mima by way of a dowry. But there is some mystery here, for Clarke's will makes no mention of dowries. Instead it provides that both his daughters would receive a share of his estate on their mother's death, which is rather a different arrangement.[27] Nevertheless, Lily's conviction clearly soured relations between Timperleys and Rolts and caused a family breach.[28]

Tom was born on 11 February 1910 at 7 South View, just off Eaton Road in Chester. He lived there only until his fifth year. One of his most potent memories of Chester was of 'dragging' his nurse to the site of the cast iron girder bridge over the River Dee which had collapsed on a spring evening in 1847 just as a train crossed it. Some seventy years after the event, local people still talked about how the carriages had plunged into the self-same river whose muddy waters Tom could see between the planks of the footwalk.[29] He also had recollections of travelling to Lancashire in the sidecar of his father's Williamson Motorcycle to visit relatives and of being hurried past the nightmarish slums of Handbridge, his nurse hissing at him to spit on the pavement to avoid breathing the stench of the place and picking up infection.[30] Around this time also, he began to form some sense of the family to which he belonged.

His Rolt relatives – his father's siblings – were a frequent presence in the background to his childhood. About his Rolt aunts – Marjory, Dorothy, Gladys and Hilda – he says little, although he tells a tale of Dorothy's making a train wait for her while she milked a goat. But his recollection of the uncles *en masse*, arguing in their high-pitched voices is keener. 'Lacking both money and training', he asserts, 'they all led fairly unsatisfactory and frustrating lives.'[31] A sweeping statement, it hides the enterprising, adventurous streak which, on his own admission, led them to seek their fortunes overseas or to follow unconventional career paths. Algernon Rolt, he relates, took up a posting in India. Henry, known as Harry, was for a time an actor and Vivian a painter. Neville and Wilfrid worked as tea planters in what is now Sri Lanka where, between 1904 and 1921, Neville managed the Labookelle estate while Wilfrid worked for the Atherfield and Yaha Ella Companies over 1898–99.[32] From 1904–1908, the brothers jointly owned the Sacombe Estate. The name 'Sacombe,' incidentally, held considerable potency to the Rolts, it being the family's seventeenth-century Hertfordshire home, which in 1735 had passed into the hands of the Caswall family. In a strange turnabout, John Rolt of *Moral Command* fame, apparently came calling at Sacombe when he was an 'unknown and penniless young lieutenant' and, one thing leading to another, ended up marrying heiress Anne Caswall.[33] It may explain why so many Rolt sons – including Tom – came to bear her family name.

Notwithstanding *Landscape with Machines*' tacit suggestion that they were gentlemen of leisure, during the 1900s the Cheshire-based brothers were in business locally and Lily, in his entry for the 1911 census, gives his occupation as 'brick maker'.[34] Indeed, the brickworks was evidently enough of a family concern for the workforce to send a wreath for Annie Rolt's funeral.[35] But the Rolt brothers' association with it did not survive the outbreak of war and by November 1914, the Hoole Bank Brick Company had been taken over by a Mr Burley.[36]

Around this time, actor Harry's marriage had recently ended and he left London to settle in Hay-on-Wye on the Welsh border. From the first, he adored the place, thinking it 'a splendid little town'. There he gave public recitations, directed amateur productions, coached private pupils for stage careers, won prizes for his roses and delighted local audiences with his conjuring displays. In 1913 he gleefully told a meeting of the local Constitutional Club how, when first he had arrived there – a 'total stranger' – he had 'met kindness at the hands of everybody'.[37] His enthusiasm was infectious. The following year, Lily left Chester and brought his wife and small son to live just outside Hay in the village of Cusop.

'Radnor View', the Rolts' new home, came onto the market early in 1914. Only recently built, the advertisement in the *Brecon County Times* described it as a 'pleasantly situated residence', with three reception rooms, four bedrooms and 'an excellent garden with large lawn'.[38] If it sounds rather a large dwelling for a married couple with one young son, accommodation was necessary for the Rolts' domestic staff. Lily and Jemima may have hoped to raise a sizeable family there although it turned out that Tom would be their only child. Solitary by inclination, in *Landscape with Machines* he admitted that he 'was, and still am, an unsociable animal' but his fondness for his own company did not get in the way of his making friends with a local farmer's son, probably one of the Lilwall family who farmed the Llydyadyway estate.[39] But for much of the time, and typically of the experience of Edwardian children, his chief companions – the adults with whom he spent the greater part of his time – were his adored nurse, Mary Gwynne, and his tutor, Thomas Southwick.[40]

About Mary, he tells us little, although a touching Christmas card that she addressed to 'Master Tom' survives among his papers.[41] He says rather more about Mr Southwick who was a retired schools inspector with a great sense of commitment to the Church and the Cusop community, and whom he remembered as a 'mild, scholarly and kindly man.'[42] In his spare time, Mr Southwick lectured on the work of the Church Missionary Society and assisted in the National Registration – in effect, an extra-ordinary census intended to form the basis of a register of the civilian population.[43] For him, Tom painted pictures – one of them shows a resplendent 0-4-6 locomotive in red and violet livery – and conscientiously wrote short 'compositions' about visits to the Taylor family in Chester, watching a local football match and going with Lily to Cheltenham which he wrote out in shaky copperplate script.[44] Although Tom remembers his tutor as an amiable man, Mr Southwick was clearly rather a martinet, and slipshod spelling brought out all his strictness. Corrections with each misspelled word written out three times over appear at the bottom of most of pages in Tom's exercise book.[45] 'Careless' is Mr Southwick's standard reproach for writing in which sentences muddle, or from which, in the surge of narrative, Tom omits crucial words or forgets to punctuate. At the same time, he was happy to indulge Tom in his fondness for Frances Hodgson Burnett's tale *The Secret Garden,* about which Tom wrote many short essays.

Cusop, in common with the entire district around Hay on Wye, was at this time a 'horse-drawn world' and, in what he came to view as a singularly 'misguided moment', Tom's parents bought him a Shetland pony. Meg, as she was named, reached Hay station in a crate, having been despatched by train from Aberdeen and her arrival provided subject matter for one of Mr Southwick's composition exercises. From its brimming details, it sounds as though the pony at first captivated Tom utterly. Meg's mane, he relates, was brown but otherwise she was 'quite black'. 'We groom her every day,' he writes:

> and Daddy went down in to Hay and bought her a stiff brush and
> a curry comb. It [sic] is so tame and lets us stroke it.[46]

Tom's excitement about the pony would not last. Released from her crate, the little mare had been at first too stiff to stand. Despite her apparent docility, the experience of travelling in cramped conditions seems to have left her suspicious of her new owners. Tom was both too young to recognise how easily she could become frightened in her new surroundings and too inexperienced to win her confidence. In another of his schoolbook writing exercises, he relates that she reared when he was riding her and then lunged at him, swinging round to kick out when he dismounted. As a result, he came to detest her, losing all interest in horses and dismissing Meg years later as a 'vicious little brute'.[47]

Lily, as it happened, shared his son's preference of wheels to hooves and, notwithstanding the prevalence of horse drawn transport around Hay at this time, relied on a pre-war 2 ¾ hp solo A.J.S. motorcycle for 'personal' transport, with a sturdy Williamson combination for family use.[48] As a purchase, Tom remarks, the Williamson bike with its substantial sidecar represented something of an aberration on Lily's part since he was, in the main, a loyal A.J.S partisan. Nevertheless, it is easy to see that it promised to lend itself to local errand-running and shopping trips. Among *Landscape with Machines'* illustrations is a photograph of a boy sitting astride the Williamson bike with an adult in the sidecar.[49] The caption offers no indication of the identity of rider and passenger, but it is likely that they are respectively Tom and Lily. With the passenger half turning towards the photographer and the boy's feet barely reaching the running board, the picture has a leisurely, summer afternoon quality, evincing a certain pride in the motorcycle, if little sense of immediate endeavour. As Tom grew older, he began to recognise the Williamson's shortcomings. Not only did it strike him as 'unwieldy' in build, but he asserts that it was less than reliable, being best kept for local journeys.

A motorcycling accident soon after the end of the First World War persuaded Lily that he needed a safer form of transport. The few cars that were available for purchase at the time commanded excessively high prices but, undaunted, Lily paid something approaching £500 for a Willys Overlander complete with 'accessories' such as headlamps and horn for which the manufacturer charged extra. An opinionated boy where cars were concerned – 'insufferable little car snob' is how he came to describe his youthful self – Tom thought it a poor bargain. A glutton for fuel, it was in his view, 'very dreary' although he grudgingly allowed that it was reliable, insofar as it could manage stiff ascents and 'go up anything in bottom gear' which was more than could be said for other cars of the time. On holiday with his parents in Pembrokeshire, at the age of about 10, he had watched drivers falter time after time on the small, steep, South Walian hills before realising that to cope with the gradient they would have to tackle the slope in reverse gear. It was, he recalled, the only time that he was inclined to feel superior about the Overlander.[50]

About the War, which must have overshadowed the Rolts' Cusop years, Tom says little except to remark that it barely touched upon 'the even tenor' of his and his parents' lives. Now in his forties, Lily was deemed to be too old for active service, but his skill as an angler and shot combined with Jemima's work in the garden ensured that the family barely noticed the effects of rationing.[51] Yet the war struck fairly close to home. Herbert Timperley, Tom's invisible half-uncle, was commissioned 2nd Lieutenant as one of a special reserve of officers and served first in the Royal Field Artillery and later in the Royal Horse Artillery.[52] That Herbert's half-sister, Jemima, should not at least have paused to consider if he was alive or dead seems improbable. While Tom, a small boy, may not often have been party to his parents' conversation, he has clear memories of his father, 'too old to serve, shaking his head over the headlines in the newspaper'. Yet, of his mother's reflections, he says nothing. Tom's abiding memory of the armistice of 11 November 1918 was of hearing the works hooter at the sawmills 'suddenly and unexpectedly' in the middle of his lessons and Mr Southwick's sitting back in his chair to observe with a sigh, 'Well, it's over at last'. It was, Tom admits, only after the event that he began fully to grasp 'the terrible carnage and agony of the Great War'. It is possible that his dawning apprehension of the catastrophe marked the origins of his pacifist convictions which his distaste for the type of militarism that he would encounter at Cheltenham College no doubt reinforced.[53]

Beginnings

Cusop Dingle, through which the Dulas Brook flowed and where trout curved through the waters while the sun shone through the overhanging hazel trees and dippers bobbed on the midstream rocks, was a constant delight.[54] In time, Tom took to exploring the surrounding country with its lonely tracks and the high plateaus and ridges leading to higher, lonelier hills. It was a place of profound delight – a place by which, as an adult, he 'learned to set [his] course'.[55] He even recalls one occasion when 'Granny Timperley' came to stay at Radnor View and Lily hired a horse-drawn wagonette to drive them all up the track that led over Hay Bluff and over the Gospel Pass to see the hidden Vale of Ewyas and Llanthony Priory.[56] Long after the 'Radnor View' years had ended, he set his short story *Cwm Garon* in the Welsh Marches where the 'shadowed face of the great massif' seems 'as unreal and as menacing as a thundercloud'.[57] It was where first he had the sense of the landscape's elemental power which would often haunt his adult writing.[58] With its surrounding hills and streams, this 'little world of the Welsh Border' held, for him, its own distinctive charge. 'Its beauty and wildness,' he writes, 'was capable of inducing in me a strange feeling of intense exaltation that was part awed reverence and part terror.' From this sensation, 'and not from any organised religion', sprang his 'conviction that there is a God' – a God, that is, whose existence far surpassed 'human conceptions of good and evil'.[59]

Chapter Two

'the dirty devices of this world'

Lily's brother Algernon – Tom's 'favourite uncle' – was, inadvertently, the agent of the Rolts' departure from their Welsh Marches paradise. Tom states that Algernon was the well-paid tutor of the heir of the Maharajah of Cooch Behar, which is not implausible. He was also, at one time, an officer of India's Court of Wards – a responsible, not to say demanding post – and in the course of his work he was accused of 'malpractice' – specifically, of receiving bribes. His superior, a Mr Lea, initiated the proceedings against him, apparently on the evidence of a servant whom Algernon had previously dismissed. The case came before the High Court of Calcutta on 14 December 1904 and by 22 December, the jury had returned a not guilty verdict and Algernon was acquitted.[1] But it proved a Pyrrhic victory. In place of his former well-paid post, Algernon was offered only a 'miserable appointment of a bare Rs300 a month in Calcutta' which he saw as a snub intended to 'deprive' him 'of all hope of getting anything better' and although he 'decided to appeal to Parliament and the public' his efforts proved fruitless.[2] What, for Algernon, seems to have been the last straw came in 1909 when he answered an advertisement for an estate manager in Purneah with a salary of Rs500–Rs1000 per month only to realise, after he had paid the recruitment agency Rs10, that no Purneah landowner could afford such a sum and that the advertisement must have been fraudulent.[3] By the summer of 1910, he was back in Christleton, the Rolts' Cheshire base.[4] Sometime later, he settled in London to pursue a 'life of slightly raffish bachelordom as a man about town', which is how Tom remembers him.[5]

Algernon, whose comfortable existence may have owed something to a taste for financial speculation, persuaded Lily to invest heavily in a company which failed. In the late summer of 1920, Tom's errant uncle arrived at Cusop, 'urbane and charming as ever', to impart the news. The sound of raised adult voices and his mother's tear-stained face told their own story and Tom guessed that there had been some sort of disaster. It was a particularly savage instance of what Lily termed 'Rolt luck' – bad luck, in other words – and it placed the family in what, from the perspective of their rather comfortable existence, rated as a major financial crisis. In pursuit of economies, Lily impulsively sold his thirsty Willys Overland and visited the 1920 Motor Show where he bought an

'the dirty devices of this world'

open two-seater Belsize-Bradshaw by way of replacement. He paid £5 on top of his new car's purchase price for a dickey seat for Tom but declined to part with a further £15 for an electric starter. Witnessing his father's later exertions with the Belsize-Bradshaw's starting handle, Tom decided that the small-scale saving was woefully misplaced.[6]

More significantly, Lily and Mima decided to sell Radnor View and look for a house which lent itself to easier upkeep so that they could dispense with domestic staff. Having taken stock, they decided to move into the vicinity of Cheltenham. Not only was it a town that they knew and liked, but the surrounding area promised to have considerably better communications than remote Cusop, or even Hay, could claim. Besides, they found the nearby Cotswold countryside appealing.[7] Their search for a Gloucestershire house that lent itself to economical living took them eventually to 'The Cottage' in the hamlet of Stanley Pontlarge which they bought when it came up for auction at the end of April 1921.[8] Originally two modest dwellings – one eighteenth century, one significantly older – which a previous owner had converted into a single residence, the auctioneers' notice described it enticingly as a 'charming, old-fashioned, stone-built country cottage residence'. Not only did it boast 'excellent ground-floor kitchen offices, new grates and modern improvements' – all details which helped to satisfy Mima that she would be able to manage the house with minimal extra help – but it was located reasonably close to Charlton Kings, the current home of her ageing mother and, although Tom does not mention him, her light-fingered clergy-stepfather.

It is possible, that it was when the Rolts purchased Stanley Pontlarge that the full extent of Isaac Timperley's 'depredations' upon Mima's inheritance came to light. Timperley's career in the Church gives some suggestion that he enjoyed comfortable, not to say, lavish means. His pastimes included playing golf and breeding horses.[9] Appointed rector of Hambledon, near Godalming, in 1908, before he moved into the Rectory, he spent £1500 on refurbishments to make it habitable.[10] In November 1914 he resigned his Hambledon living on account of ill-health to embark on a long, nomadic retirement, during which Mary and he would reside in Weston-Super-Mare and Ascot before settling at Charlton Kings. There they spent their final days.[11] The couple died within two months of one another – Mary late in January 1928, her husband on 19 March of the same year.[12] In a suggestive twist to the tale, it emerged after Isaac Timperley's death that he had left almost £15,000 to his son, Herbert. Except for the gift of £25 to a servant who cared for him in his final weeks, it was his sole bequest.[13]

Pleased with what was to become their new home, the Rolts moved in sometime between the late spring and early summer of 1921. Mima decided that the occasional tables, fire-irons and languorous art-nouveau vases which had suited the design of Radnor View did not chime so well with the Cottage and she embarked on a scheme of transformation. One of her ideas was to remove the patterned wallpaper so as to leave the walls plain, under a wash of cream or white which highlighted the colour of chintz curtains and broad bowls of flowers. Lily saw the trophies of his cycle-racing days consigned, ruthlessly, to the attic with other pieces of 'Edwardian bric-à-brac', while his gun-case vanished from sight under the stairs.[14] As for Tom, the experience of moving house unsettled him and he could muster little liking for his new surroundings. After Cusop, in sight of the great hills of the Welsh border over which the changing light and shadow played, on whose expanses whinberries and cranberries grew in profusion every summer and in whose folds the winter snow lingered, the Cotswold pastureland struck him as 'tame.' A pathway near the Cottage led up Stanley Mount, a high point on the edge of the Cotswolds from which, tantalisingly, he could discern the profile of the Black Mountains on the western skyline and would try to console himself with the thought that they were not so very far away. Once he grew familiar with the pattern of the farming year, he developed some affection for the nearby fields and orchards with their tidy stone walls, but he insisted that they 'remained always very much a second love'.[15]

More certain comfort came from the railway which ran past the garden. By this time, Tom was a seasoned railway enthusiast. Hay Station had served both the Midland Railway Company's Hereford, Hay and Brecon line and the line which the Golden Valley Railway Company built from Hay to Pontrilas which the GWR acquired in 1901, but in traffic terms it was rather dull. Each railway owed its existence to lavish nineteenth-century ambitions which never fulfilled their promoters' expectations. The Hereford, Hay and Brecon branch had originated in Thomas Savin's plans for a route by which to transport iron ore from the Midlands to Ebbw Vale, but these early ambitions were never to be realised. When the Rolts were living in Cusop, it was an 'isolated backwater' over which ran trains of four-wheeled carriages lit by pot lamps, in which 'brilliantly hideous' carpet-covered footwarmers took the place of steam heating. As for the Pontrilas branch, once mooted as part of a connection between Bristol and Liverpool, it was evidently a law unto itself. Tom remembers it as running three 'mixed trains of passengers and goods daily in each direction' while keeping firmly to the 'one engine only in steam' principle but making no perceptible attempt to keep to anything resembling a timetable. Hoping to visit Pontrilas one afternoon, Tom and his mother spent such a long time watching while the small green tank engine shunted in Hay goods yard, 'every movement punctuated by a prolonged pause' that they eventually lost patience and abandoned all hope of making their journey.[16]

Chester held greater promise and from the nursery windows of Grey Friars House, home of his Taylor relatives, he could see trains as they crossed the 'long viaduct of squat brick arches' which carried the railway across the tract of land known as the Roodee. Watching enthralled through binoculars he marked the liveries of the different railway companies, 'the gleaming black locomotives of the North Western contrasting with the more familiar green ones of the Great Western' and the 'strange coaches with salmon pink upper panels' of the London and South Western through trains, heading 'between Chester and the South Coast'. Once he was old enough, he would take the tram to Chester General Station to admire the expresses. The Euston-Holyhead express trains usually arrived double-headed – 'a Precedent piloting a George V or a Precursor' – and hearing the distinctive reverberation of the two locomotives' coupling rods re-echoing from the station roof, Tom at the time thought it nothing less than the 'proud voice of power'; a uniquely 'brave and thrilling sound'. Only later did he come to suspect that it 'may have been a symptom of technical imperfection'.[17]

Visits to Plas-y-Garth, the Taylors' house in North Wales, brought him into the ambit of the Glyn Valley Tramway, for whose Beyer Peacock 0-4-2 tank engines *Sir Theodore*, *Dennis* and *Glyn* he held in deep affection. In *Lines of Character* (1952) – his survey of what the book's blurb terms the 'remote and lesser known outposts of the railway systems' – he recalls how he used to hope that *Sir Theodore* would head his family's train. 'She was my favourite', writes Tom, who belonged to a generation of railwaymen which unhesitatingly assigned female pronouns to their locomotives, even when they bore male-appellations. The name 'Sir Theodore' caught at his sense of romance, in a way that the 'prosaic' 'Dennis' or 'Glyn' never would. Locomotives' names counted for much with him in his boyhood and to travel behind George Jackson Churchward's Star class locomotive *Knight of the Grand Cross* invested an entire journey 'with the magic of a crusade'.[18] When he was about 12, the Glyn Valley line gave him one of the great experiences of his early life. Seeing him linger opportunistically by the water tank at Glyn Ceiriog, one of the drivers read his thoughts and offered him a footplate ride. Admittedly, the locomotive was not *Sir Theodore*, or indeed any of the original tram engines of the Glyn Valley fleet, but an ex-Rail Operating Division Baldwin 4-6-0 tank engine – a make notorious for giving a rough ride – yet he travelled in triumph, 'speechless with gratitude and wonderment' amid the heat and noise up to the quarries at Hendre, to return with a load of roadstone.[19] It was one of his life's landmarks.

The move to Gloucestershire brought him onto long-established GWR territory and from a comfortable point of vantage within one of the Stanley Pontlarge apple trees, he would admire the elegant design of the locomotives of the Atbara and Flower classes as they hauled express trains bound for the Midlands. Even more thrilling were the express trains that ran between Wolverhampton and Penzance 'once a day in each direction'.[20] The Cornishman, as the train came to be known, may have been Tom's inspiration for some exuberant verse. 'The glorious sun is sinking to rest,' one of his youthful poems opens:

> And the metals like bars of gold
> Stretch, narrowing, far mid the purple haze
> Into the west.
>
> The Blood-red orb can be seen no more
> But only a crimson streak,
> As out of the dusk I see her come
> With a gathering roar.
>
> And now I see her before me loom
> And now she's hurtled passed [sic]
> With a flash of vermilion, green and brown
> Out of the gloom.
>
> The glorious sun has sunk to rest,
> And the light like a scarlet star
> Is fading away as the sun has set
> Into the west.

While the prosody and sentiments may not be especially memorable, Tom's observation of the sunset colours – gold, crimson, vermilion and scarlet – is keen and there is no doubting his gusto. Another poem runs:

> Away to the west
> On the Line that is Best
> O this is the life for me!
> With the hiss of steam
> And the whistle's scream
> As I thunder by.
>
> Away to the west,
> To the wild cliff's crest
> O this is the life for me!
> With the line all clear
> At the junction near,
> As I thunder by.
>
> Now far in the west
> On the Moorland's breast,
> O this is the life for me!
> Mid the murmur of brake
> And the sunset's wake,
> I glide to rest.

The lightest of verse, Tom's expression of the locomotive's thinking speaks of his own feelings at seeing it pass. Precisely when he composed the poems is impossible to say; the *Landscape* books make no mention of them and the manuscript in which they appear bears no date.[21] But it is possible that they had their origins in the summer Saturdays that Tom spent in the fork of an apple tree overlooking the GWR line as it passed the Stanley Pontlarge orchard.

Not long after the Rolts' move, an ugly scandal blew up concerning one of their neighbours in Cusop, namely Major Herbert Armstrong. A solicitor by profession with a practice in Hay and prominent among the congregation of Cusop Parish Church, Armstrong, spruce in his major's uniform with his wife at his side, often used to join the Rolts for walks and picnics. Tom had viewed the major with slightly nervous awe and much preferred Mr Southwick as a companion. Nevertheless the news that Armstrong had been arrested and charged with his wife's murder was a major shock.[22] Tom had been on the train with his mother at the time after a day in Cheltenham; she had bought an evening paper and found herself confronted with the lurid story.

Mrs Armstrong, it emerged, was not her husband's only intended victim. He had also attempted to despatch one Oswald Martin, the solicitor who had purchased 'Radnor View' from the Rolts. His trial, which took place in April 1922, caught the attention of the nation, but what gave the matter particular interest to the Gloucestershire press was the detail that Martin's family came from Tewkesbury; they had, therefore, a Gloucestershire connection.[23] Although Armstrong insisted that he was innocent, the jury returned a guilty verdict; his appeal was dismissed and, still protesting his innocence, on 31 May 1922 he was hanged.[24]

Tom's parents were heartily relieved that they were not required to give evidence in the course of the proceedings.[25] Armstrong had after all been one of their circle, if not precisely a family friend. But Tom's feelings were rather different. The incident caught his curiosity leading him to probe his memories for indications of Major Armstrong's more sinister tendencies. In fact, Tom's clearest recollection of Armstrong was of watching him wind up the carbide container which held the lighting plant that he rigged up in an outhouse in his garden. Not in any way ominous in itself, he wondered whether the memory owed something to the similarity between the 'white powder of the spent carbide' and the arsenic with which Armstrong had poisoned his wife, attempted to poison Martin and which he claimed to have bought in order to make weedkiller.[26] But the image of Armstrong hauling up his lighting apparatus stayed sharp enough in his thinking for him to 'hear the clicking of the pawl on its ratchet'. On one level, the episode is no more than a rather gruesome piece of history in the background to his childhood. At the same time, Tom's deliberate quest for the telling detail – some revelation of Armstrong's criminal intent – shows something of the writer-in-the-making, consciously seeking out the workings of cause and effect. It was a skill that he would, in later life, use to good effect.[27]

Lily and Mima were determined that their son should receive a public school education and since Lily's brothers Neville and Wilfred had both been educated at Cheltenham College and Vivian the artist had sent his John there, it struck them as the obvious school for Tom. Besides they liked Cheltenham with its useful shops and elegant architecture and often came to Sunday services in the College's imposing chapel. Whether, before they despatched him to the preparatory department, they paused to consider how readily their solitary son who loved roaming the hills but 'always disliked children's parties' would adjust to living among other boys within the confines of dormitory, classroom, common room, chapel and playing fields seems unlikely.[28]

Tom viewed his despatch to boarding school in the light of Adam's expulsion from Eden.[29] Cheltenham College Junior School, he allowed with adult magnanimity, was probably 'no worse than other preparatory schools of the period and probably a great deal better than some', while in the same breath recalling his boyish incredulity that the masters 'could have created such a barbarous little world'.[30] It brought him into contact with what Thomas Traherne – later, one of his favourite poets – termed the 'dirty devices of this world', and he describes his first term as 'traumatic', a word which, in its literal sense, suggests he found the experience of school life completely bruising.[31] The experiences that he describes do not, to be honest, sound brutal in the *Tom Brown's Schooldays* sense and *Landscape with Machines* makes no mention of either of fagging or bullying, for instance. Nevertheless, Tom found Cheltenham College Junior School quite unlike anything that he had ever come across.

From his description, the school seems to have made the tacit assumption that its boys were all criminals in the making. It confined them within the narrow bounds of classroom, dormitory, playing fields and yard, with compulsory daily attendance of the College Chapel and a walk under the surveillance of a master on Sunday afternoons.[32] In the dormitories, iron nails bristled along the pitch-pine walls of the cubicles in which the boys slept 'to discourage climbers'. Each day's timetable followed the unspoken expectation that 'if every waking minute of work or play were not ordered, dragooned and disciplined' the boys – louts and ruffians, all of them – 'would run riot and commit fearful crimes'. For the most part, masters and boys alike lived a in crowd, moving *en masse* between classroom, common-room, refectory and dormitory; some of them, Tom concedes, even enjoyed it. But finding himself adrift upon the currents of 'hearty communal life' quite unlike anything that he had ever known previously, he sank into melancholy. In the long light evenings, he observed the unknown family who lived in one of the houses near the school, his thoughts turning all unbidden to his father playing bowls with Mr Southwick on the Radnor View lawn as he saw the strangers sit in their garden under a cedar tree. Even when darkness fell, he lay awake 'thinking by the hour [...] of Cusop Dingle, of the Black Mountains and Llanthony'. Untrammelled misery took its toll and before long, his homesickness turned into a fever at which point the school contacted Lily and Mima and asked them to take him home.

Although it was no more than temporary, his respite at least gave him an extra-long holiday. Yet Tom at this time excelled academically. He won a prize for Geography in his first term and throughout his first year came either first or second in his class for English.[33] To attempt to diagnose the source of his future trouble at school about a hundred years' after the event would be foolish, but it may have owed less to Cheltenham's tough, hearty regime than it did to the anxiety that Tom may have had concerning the effects of the 'misfortune' had upon his parents, combined with his own unhappiness over leaving Cusop. The increasingly pointed term-by-term comments that the head of Cheltenham's Junior School – Basil Allcott Bowers – wrote on his Report Card suggest that Tom's engagement with his school work declined steadily over his first four terms. At the end of the spring term of 1922, Bowers judged Tom's performance to be 'satisfactory'; by the end of the following term he remarks that Tom 'obviously takes a rest now and again' – was apt, in other words, to allow himself to coast and fall back. In December 1922, when Tom had been in Cheltenham's Junior House for nearly a year, his criticism sharpened. 'Good', Bowers concedes, before adding 'but could not some holiday measures be taken to improve his French?' A term later, he finds Tom's lack of prowess frankly aggravating. 'Not satisfactory', he pronounces, 'and that French remains a blot.'[34]

Tom missed the summer term of 1923 entirely. For this large gap, his Cheltenham record supplies no explanation, but the cause was probably connected with his health; apparently, he had been so sickly a child, that at one time his parents feared that he might not survive beyond infancy.[35] In *Landscape with Machines* Tom dates his long illness to 1920, but indicates only that the school doctor sent him home before the term's end. It is possible either that his memory deceived him or that during his time at Cheltenham, he had more than one extended period of absence. Interestingly, he mentions 'having grown another skin' in the course of recovering from illness before he joined the Senior School. It lends substance to the idea that he toughened up enough, at least outwardly, to view his return to the Cheltenham fold if not eagerly, then at least with 'with stoical resignation'.[36]

For the following term, his record carries the encouraging comment 'a useful recovery from illness' and Tom came top in his set for English and a creditable third in Science.[37] Despite this success he never conformed to Cheltenham College's expectations. Like other public schools, Cheltenham prized athleticism and sociability, but Tom was neither sporty nor gregarious. The minute organisation of the school day so as to ensure that no boy had any time to himself, he describes as frank 'torment'.[38] Eventually, he found a couple of kindred spirits, boys who shared his impatience with the college's finicky regulations and traditions, and the three of them made up what was, no doubt, a useful alliance against the more egregious manifestations of 'college spirit'. One this trio of malcontents was Harry Rose who, years later, would help him to secure a sustaining wartime job in the Ministry of Supply.[39]

Even the friendship of like-minded boys did not, in Tom's thinking, make up for the inadequacies of the curriculum. Cheltenham College had placed him on what it termed the 'Military and Engineering Side'. Unlike the boys who went on the Classical Side to focus on Latin, Greek and Ancient History, it meant that his timetabled lessons were in Mathematics, English, French (with which he always struggled), Science, Drawing and Geometric Drawing. These subjects were perhaps not as clear-cut as the list suggests. Under the heading 'English Syllabus', for instance, the College listed as required reading not only *Hamlet* and the General Prologue to *The Canterbury Tales*, but also Warner and Marten's *English History*, Burrell Smith's *European History* and Unstead and Taylor's *Essentials of World Geography*.[40] To all appearance, it sounds like a ragbag combination of various humanities disciplines that had no connection with Ancient Greece and Rome. 'Geometric Drawing' probably meant technical or engineering drawing, while 'Drawing' was closer to art. Tom, who drew well, tended to be significantly higher placed for 'Drawing' than he was for 'Geom. Drawing'.

T.C. Currie, Tom's form master, described his Maths at the end of his first term in the Senior School as 'v'poor'; on the other hand, his English was strong and Currie reckoned that as a new member of the Military and Engineering Side of the College, Tom had made a fair start. 'Pretty good for his first term as a Senior' approved Major Henry Hardy the headmaster, noting that the boy's ambition was 'To be an engineer', which would have been Tom's reply to any question about his choice of future career.[41]

By and large, the overall record of his form placings suggests that Tom continued to do well in English throughout his time at Cheltenham, but lost ground in Mathematics, Science and Geometric Drawing where he found himself placed increasingly near the bottom of his class. It was his misfortune that at Cheltenham, success in English and Drawing cut little ice. 'All artistic activities', Tom recalled years later, 'were considered "cissy" and to indulge in them was ... almost as shameful as masturbation.'[42] 'Doing quite well', wrote Currie cautiously in summer 1924 and the following term Hardy pronounced him 'Well up for his age'. But then something went wrong. In spring 1925, not only did Currie castigate his 'v'bad French' but Hardy witheringly observed, 'Not very grand, is it?' which reads like an allusion to some episode on which Tom preferred not to descant in his autobiography. After that, if it was not quite downhill all the way – something briefly moved Hardy to observe that 'the outlook improves' while he was 'delighted' to hear of Tom's 'better endeavours' – Tom's academic progress was hardly promising. 'Casual' pronounced Currie damningly at the end of 1925; 'completely self-satisfied.' A term later, 'Maths fearful' he reported.[43]

By this time Tom had lost patience with his educators. In *Landscape with Machines*, he states that his especial grievance was the College's complete failure to teach him anything about engineering.[44] 'Geom, Drawing' – the discipline on the College curriculum which most closely approximated to a practical skill – was not a subject in which he shone. He came fourteenth in a class of seventeen in spring 1925; twelfth out of sixteen in summer of the same year, and thirteenth out of seventeen at Christmas, when he compounded the ignominy by coming bottom of his set in Mathematics.[45] Resolute, he seized the initiative and asked Lily to allow him to leave and, to his surprise, Lily acceded to his request.

Matters came to a head in April 1926 when Hardy interviewed father and son together in his study. For evidence of what passed between them, we have only Tom's account in *Landscape with Machines*, which he wrote long afterwards, when other events may have clouded his memory. According to his recollection, Hardy, resting his elbows on the surface of his vast desk and putting his fingertips together, leant forward and sought to persuade Lily Rolt of the folly of removing Tom from Cheltenham College.

'I suppose you realise', he said, 'that by taking this step, you are ruining your son's career?'

'That', Lily replied, 'is a matter for me to judge' and before Hardy could say anything more, he rose, linked arms with his son and led him away from the narrow world of school for ever.

Whether Tom's exodus from Cheltenham really worked out in this way we do not know, but his departure assuredly gave Hardy reason to reflect. 'Another year', the headmaster wrote on Tom's record card, 'would, we think, have been valuable to him, though perhaps not technically. I regret his early departure. [He] leaves for a machine shop at Evesham'.[46] To Major Hardy, who was accustomed to despatching appreciative Cheltonians either to Sandhurst to become army officers or to Oxford or Cambridge before they joined the Colonial Service, it must have been singularly disappointing.

Chapter Three

The Apprentice in the Country

The 'machine shop' was in fact the modest agricultural engineering business which the Bomford family ran from the village of Pitchill on Evesham's fringes. There, Tom's 'courtesy uncle' Kyrle Willans, husband of his godmother Hero Taylor, found him a pupillage – an unpaid post which promised to give him some introduction to the engineering profession.[1] Willans was an enigmatic, unpredictable presence in Tom's life and, it is probably fair to say, in the lives of his employers and work colleagues. Chief engineer to a small Northampton engineering firm, he followed in the footsteps of his father Peter Willans who had established a ship building company at the Ferry Works, Thames Ditton; invented an extremely efficient high-speed steam engine and was the recipient of the Institution of Civil Engineers' Telford Medal.[2] He died at the early age of 41, thrown from his dog-cart on his way to catch a train.

Quite unlike the Rolt relatives with their aimless lives and interminable reminiscences, Kyrle Willans gave Tom an early taste of the pleasure to be had from driving a car – specifically a 1906 Rover – which Tom had proudly piloted 'round and round' the yard of the Willans family's Milton Malsor home. He was, beyond question, the obvious person to turn for advice after Tom's abrupt departure from Cheltenham College. The Bomford family were friends of his. Not only did their work include 'STEAM PLOUGHING & CULTIVATING […] REPAIRS TO ALL CLASSES OF MACHINERY, with 'TREE PULLING & STUMP GRUBBING and BOILER WORK A SPECIALITY' which promised a variety of experience, but Pitchill was no great distance from Stanley Pontlarge.[3] If Tom's parents were anxious about him, it may have been reassuring to reflect that he could easily cover the fifteen or so miles on his recently acquired 2¼ hp B.S.A motorcycle.

Steam powered machinery had become increasing prevalent within agriculture since about the mid-1850s and paterfamilias Benjamin Bomford (1828-1880) – according to tradition, 'the greatest arable farmer that Worcestershire had ever produced' – had championed its use. His sons Benjamin and Raymond inherited both their father's skills and his interest in agricultural engineering and after his death, they formed a commercial partnership known as R & B Bomford.[4] By the time that Tom arrived at Pitchill, management both of the farm and the contracting business had passed to Raymond Bomford's sons. Of the new generation, one son – the younger Benjamin Bomford – left the partnership at an early stage to concentrate on fruit farming and another – Richard – eschewed farming for medicine. But Douglas, the eldest brother, directed the mechanical and contracting side of the concern from Bevington Hall, the old Bomford family home, while younger brothers Leslie and Ernest, known as 'Hercs' on account of his herculean strength, ran the farm from Pitchill House.[5] Sometime after Tom's arrival Leslie, whose service in the Great War had earned him a Military Cross and a Distinguished Service Order, moved to Shrewsbury leaving Hercs to manage the farm on his own. Tom lodged with him in the farmhouse throughout his time at Pitchill and remembered him as 'always very friendly and jovial'.[6]

Despite having made few friends at Cheltenham, among the men with whom he worked Tom quickly found his feet. Bruce Bomford, in his family history, states that in the 1920s the company included 'two smiths, one boiler expert, a fitter who could face a slide valve and seat, a turner for the lathes, and two carpenters'.[7] Rather more specific, Tom names Percy Lester, the works foreman; fitters and machinists Will Salisbury and Alan Bloxham; carpenter Ernest Cole; Sentinel Waggon man and occasional steam plough driver Will Robbins; steam plough driver Bill Smith, who took a memorably lax approach to basic safety precautions; Archie Ellison in charge of the McLaren ploughing engines; one-armed Joe Bailey who looked after the portable engine that powered the workshop; George Leonard the shop boy; and a blacksmith whose name he cannot recall – a total of ten onsite staff. From them, he began to 'learn the rudiments of the engineer's craft'.[8]

Seeing them at their day-to day tasks, he assumed at first that much of the work was undemanding. Observing the engineman Joe Bailey, for instance, and marking how he 'never hurried' but 'occasionally [threw] on a shovel full of coal, or … casually [supplied] the boiler with water' Tom guessed that he must have a very easy time.[9] Only when he had to cover for Bailey and soon found his steam and water gauges veering dramatically back and forth did he realise just how much close attention the engine actually required. The nonchalance with which Mr Bailey went about his work concealed skill born of long experience. Asked to stand in for Alan Bloxham, in charge of the mill which ground meal to feed the Pitchill cattle, the lesson repeated itself. Used to the sight of Mr Bloxham sunning himself in the mill doorway, Tom supposed that he had the most leisurely of occupations. Only through having himself to top up the grain hoppers, remove and replace the meal-sacks as they filled, keeping a watch on the Tangye engine all the while to ensure that it did not overheat, did he recognise the reality of the task. It left him rushing up and down the mill staircase like 'a kind of demented sorcerer's apprentice'.[10] Together, the experiences served to disabuse him of any inclination to assume that manual labour equated with unskilled labour. It was one of the core lessons of his time at Pitchill.

Since commercial contract ploughing was one of the staples of the Bomford business, Tom soon experienced it at first hand. The system was that two ploughing engines, one on either side of the field, drew the plough across it on a steel cable. Each end of the cable was attached to winch drums beneath the engines' boilers. The Pitchill business had, Tom relates, 'five sets of steam cable ploughing tackle, each set consisting of a pair of engines' – one right and one left-handed, so that one delivered cable from one side of the winch drum and the other from the other – and a plough, constructed to tilt see-saw wise at each furrow's end so that first one set of ploughshares cut through the soil and then the other.[11] Not only did the machinery require regular service and repair work, but travelling from farm to farm across the Vale of Evesham drew him straight into the heart of agriculture. He came to love riding the ploughs – that is, sitting above the tail of the plough frame to keep it stable – and likened it to 'sailing a ship through the earth', his only concern being lest the cable should break, recoil and decapitate him.[12] The cultivator, used for breaking up the ground and removing weeds, would be equipped with up to sixteen tines set in a triangular frame. Like the plough, it ran between engines standing at either side of the field, each one pulling it in turn. When the turns changed, the steersman pulled a lever which elevated the frame so that the tines were clear of the ground and a ratchet and pawl held them in position. As soon as the next pull started, the cultivator pivoted it on its central axle, the steersman released his lever and the tines dropped.

Besides giving him some knowledge of farm work, taking his hand on the plough or cultivator team built up Tom's native resilience. A new boy, he was ripe for teasing and turning the cultivator was the perfect opportunity. Unless the engineman began his pull very cautiously, the cultivator frame, Tom relates 'would tilt alarmingly'. 'The old steam plough drivers', he soon realised:

> could judge to a nicety how far to make it tip without actually turning over and this was their favourite trick whenever there was a new man on the cultivator. As the tyro jumped from his seat and ran for his life [...] they would shake with laughter.[13]

Tom, apparently deciding that being a butt for the regular drivers' humour was in effect an integral part of his pupillage, took it in good part which was just as well because the ribbing continued even when he progressed from riding the plough or cultivator to steering the engines. Standing on the minute metal platform outrigged from the coal bunker and 'grasping the huge horizontal iron steering wheel' was a significant test of nerve. When driver Will Robbins set off across a ploughed field 'full tilt at about 3 m.p.h.' and the axles rising and falling with the swell of the ridge and furrow, the vast wheel spun back and forth, out of Tom's control and almost flung him off his perch. Only when he realised that the steering column was equipped with a form of stabiliser did things improve. Setting his direction by tightening the stabiliser as far as he could, he left the wheel alone unless and until it was necessary to alter his course. Mr Robbins, 'a tall lugubrious man of about sixty with the expression of a sad spaniel' derived, Tom suspected, much quiet amusement from tossing his assistant about 'like a small monkey on the top of a very tall stick'.[14]

For a ploughing contract the Bomfords would despatch: a pair of ploughing engines; balance plough complete with 'two sets of five or six plough bodies' (the 'business part' of the implement, which bolted onto the frame); accommodation van; water-cart and cultivator. Besides the machinery, the work required a team of four men – one man to each engine and two men to the plough – a ploughman to steer it along the furrow and his mate who sat in a minute seat above the tail of the plough frame to keep it stable. When the plough came to the end of a furrow, Tom explains:

> its crew jumped off their seats, tipping the plough to bring the other set of plough bodies into action and, swinging it into its new position as the other engine began slowly to pull.[15]

While the contract ran its course, the team often included a fifth man – usually a labourer from the client farm – who was tasked with supplying coal and water for the engines. Tom often had to set out on his motorcycle to take messages or perhaps crucial spare parts to the Pitchill men who were out on contract. Although they might be almost anywhere in the Vale of Evesham, from the sight of steam and the sound of the engines, finding them was seldom difficult.[16]

On the road, the ploughing set was something of an obstruction to other road users, particularly to those who were in a hurry. Moving between jobs called for careful advance planning since the engines' sheer weight necessitated avoiding some of the weaker local bridges. In the opinion of the Bomford workforce, the brass plate on each smokebox inscribed '14T' – that is, fourteen tons – was utter deceit on the part of manufacturers. In the heat of summer, the strakes – ridges protruding from the engines' driving wheels – were apt to gouge deep indentations within the melting tarmac road surface, much to the frustration of the county highways surveyor. Travelling between sites had to be done in the early mornings.[17]

In winter, much of Pitchill's work consisted of routine machine maintenance.[18] Tasks like descaling boiler tubes, for instance, and patching or replacing fireboxes enabled Tom to build up a solid stock of basic engineering workshop skills – the raw stuff of Kyrle Willans' lodestar conviction that 'a pinch of practice is worth a pound of theory'.[19] He learned how to use hammer and chisel, to file so as make a flat surface and to scrape and fit bearing brasses – a bearing is the element of a machine that reduces friction between its moving parts – and the scraping was essential for achieving an accurate fit between two curved surfaces, for instance, between a cylinder and a bearing, or a bearing and a shaft. He learned to strike for the blacksmith – that is, to use the sledgehammer to do the heavy work at the smith's direction; to turn screws on a lathe; to close rivets or stays by hand which was a laborious task in the absence of power tools, and then to remove old rivets by decapitating them with a chisel bar and sledge hammer which called for a good eye and deft hand. It was a job that he thoroughly disliked. Were he, he reasoned, to lose his nerve 'and fail to hold the bar true and steady' there was a risk of the sledgehammer man's missing the mark but breaking his – Tom's – arm instead. Any work which had to be done inside a boiler or water-tank – scaling plates, for instance, or fixing new rivets – tended to fall either to Tom or to the 'boy' George Leonard because, alone among the workforce, they were slim enough to get inside the enclosed space. Besides his slight frame, being ambidextrous – 'amphibious' as one of his colleagues expressed it – gave him considerable advantage when a job required wielding a hammer in a confined space.[20] Unsurprisingly, he makes no secret of the 'bruises and bloody knuckles' that he endured at this time.

If any single incident from his time at Pitchill loomed larger than others in his memory it was one which occurred early in his time there. It did not concern skill so much as safety, for it was a striking instance of the ease with which familiarity with routine might pave the way for outright danger. Not long after he had started to work for the Bomfords, Tom helped traction engine man Bill Smith pull up some superannuated cherry trees. It meant securing a chain around the lower part of each tree trunk and hitching it to a wire rope, itself attached to one of the Fowler engines. Then Mr Smith on the footplate would open his regulator to power forward, with luck hauling the tree out of the ground by its roots. With the more stubborn trees, the old driver took drastic measures. To apply irresistible force to the immoveable object, he would screw down the wing nut which raised the level at which the Salter safety valve cut in, and allow the steam pressure to build up steadily. When he judged that it had reached the right level, he would tackle the roots again. Like any well-informed railway enthusiast of the time, Tom knew that steam needed to be treated with respect and recognised the potential perils of Mr Smith's strategy. While the pressure in the Fowler engine's elderly boiler rose higher and higher, he deliberately took cover at a safe distance. If Mr Smith, as he drew on his broken briar pipe, reckoned that Tom was making a fuss about nothing, Tom thought it miraculous that the old man avoided blowing himself to blazes. Years later, when Tom came to write *Red for Danger*, memories of that afternoon among the cherry trees rose with stark clarity in his thinking and with an insight born of experience, he would remark pointedly on the 'temptation' which a 'handy nut' might present to the crew of an underpowered locomotive facing a steep incline with a heavy train.[21]

Assuredly, the jobs that he did at Pitchill meant coming to terms with some of the filthier aspects of agricultural life and sloughing off any finicky fastidiousness with which his refined upbringing and sheltered public school life may have left him. Besides the engineering business, the Bomfords were also successful fruit farmers, their strawberry plants selling as far away as Lincolnshire. One of Tom's tasks was to collect truckloads of shoddy – woollen waste – from Salford Priors station for use as fertiliser.[22] Rich in nitrogen, the shoddy was foul-smelling and dusty; not surprisingly, Tom found it horrible to handle. Unimpressed by his complaints Will Robbins, with whom he loaded it onto the Pitchill Sentinel waggon, only commented that blood – the dried blood from abattoirs which was also useful for crop propagation – was even worse. Transporting strawberries to one of the local jam factories might have been a pleasanter errand, had not Tom happened to notice a labourer load sugar into the boiling vats on what appeared to be the boiler-house coal shovel; for some months afterwards, he could not bring himself to eat strawberry jam. Revulsion tinges his memories of the 'Dudleys' – the collective name given to the Black Country families who came each summer for fruit picking. If the contempt with which the Bomfords' neighbours viewed the incomers – 'wild, dirty and of very questionable habits' – came down to snobbery, the fleas that rose in swarms after the fruit pickers' departure in the lofts where they had slept gave rise to other objections. Rueful, Tom recalls 'cleaning up and fumigating [...] after the Dudleys' annual visit was [...] the most unenviable job on the farm'.[23] Yet for sheer squalid misery, the business of attaching and detaching heavy iron clips known as 'spuds' in the rain over the engines' smooth rear wheels marked a particularly low point. Intended to give extra 'grip' in muddy conditions, each spud weighed around a quarter of a hundred weight – 28 lbs – and when not in use, hung on a bar by the water tank. In wet weather on soft ground, they came into their own, although fitting the heavy objects – clasping one end of the spud round the wheel rim and securing it with a bolt through a hole in the wheel's surface – was a bleak job, particularly in a downpour. Taking them off again was even worse, but they were not suitable for driving on road surfaces, so the team had no choice but to unbolt the iron lugs from the mud-caked wheels in the saturated field before finishing the day's work. Tom, with hindsight, remembered it among the most disagreeable tasks he ever encountered.[24]

Not surprisingly, he viewed Douglas Bomford's experimental design of a new lightweight ploughing engine with considerable interest. Not only would it render the spuds superfluous, but it promised greatly to simplify the business of getting around. In a monograph entitled *Corn in England*, Douglas Bomford explored the efficiency of different methods of cultivation and the application of power to the land. While he saw 'steam cable engines and mechanical tractors' in widespread use, it was strikingly clear that they had not superseded the horse. The reason was that the machines' 'excessive weight and unwieldiness' meant that they were suitable only for use in dry conditions and then only for a small number of tasks. In consequence, he observed, farmers owning 'tractors and steam cable engines' had, out of necessity, to incur the cost of keeping horses as well 'to work on the many occasions when conditions are unsuitable' for reliance on machinery. At a time of high labour costs, deployment of men to drive horses hardly made economic sense; one man, Bomford remarked, 'rarely, if ever, controls more than three horsepower'. The obvious solution, in his view, lay in the development of an engine which was suitable for cable ploughing – it always promised to cause less damage to the ground than direct ploughing – but at the same time was 'light enough to travel so freely over the stubbles as to enable it to perform [...] every wholesale operation at present performed by horses'. Such tasks would include 'binding, reaping, carrying crops off the land and farmyard manure onto the land'.[25]

Tom played a part both in the construction and the early field-tests of Douglas Bomford's lightweight steam ploughing engine. Made up from parts which came respectively from Fowler traction engines (the winding drum and wheels) and Sentinel tractors (the boiler, engine, and water tank) it was, in his view, 'hardly a thing of beauty'.[26] Rather blockish in appearance, the prototype had a trunk-like water tank at one end, a stubby chimney and boiler at the other, a vertical winding drum, broad wheels and the words 'Bomford Bros., Pitchill, Evesham' along the frame.[27] Instructed to use it for cable ploughing a large, flat stubble field, Tom had the sense that Bill Smith – his opposite number, driving a heavy Fowler ploughing engine – was in ruthless mood. Having to feel his way with the prototype, as he went along he tried to work out how to get the best out of the new machine. Pulling the plough at speed, he soon realised that it disabled the water pump and set up so much vibration that he could not use the injector. It made matters fraught on a long pull and Tom worried lest he should 'run the plug' – that is, allow the water-level in the boiler to sink so low as to melt the lead plug inserted as a safety measure in the crown sheet of the firebox, which would have the effect of releasing steam and water into the firebox and putting the fire out.[28] It would not be a disaster on the scale of a boiler explosion, but would be humiliating nevertheless and Bill Smith assuredly would never forget it. At the same time, the vibrations sent all the coal to the perimeter edge of the cone-shaped firebox, leaving a central gap into which the draught mechanism sent blasts of cold air to the accompaniment of what Tom remembered as a 'penetrating humming sound'. Anyone in the neighbourhood who knew anything about steam engines would at once realise that he had made a hole in his fire and could not raise steam.

The whole experience, in his view, was 'shaming'.[29] It hardly needed Bill Smith to whip the plough across the field, his Fowler trailing steamy clouds of glory, to indicate that the lightweight engine was very much the inferior partner in the exercise. Nothing daunted, Douglas Bomford still reckoned that his brainchild showed enough promise to justify construction of a second model to partner the original. Together, Tom relates, the pair 'gave good service on contract work' until the 1930s when the Bomfords decided collectively to give up cable ploughing in favour of direct ploughing using oil-powered Fowler Gyrotillers for the work.[30]

Douglas Bomford's lightweight ploughing engine was only one of the family's innovations. Leslie Bomford – the brother honoured for his distinguished war-service – owned a Stanley steam car which Tom helped him to overhaul and quickly recognised just what powerful instinct for mechanics and 'power of invention' Leslie possessed.[31] He was not alone in marking and respecting Leslie's skills. At some point in the second half of the 1920s, Kyrle Willans joined the Sentinel Waggon Works Ltd, moving his family to Dovaston House outside Shrewsbury. Much to Tom's private regret, at the same time, he found an opening for Leslie at Sentinel's Shrewsbury base.[32]

At Sentinel's, Leslie helped to develop the steam-powered Sentinel Rhinoceros Tractor which, it is probably fair to say, was almost a commercial success. Specifically conceived for direct ploughing, the Rhino had many strengths. Its combination of rubber driving wheels at the front for ease of steering with continuous track propulsion at the rear for grip equipped it for work on rough terrain while water tanks both at front and rear gave it generous capacity.[33] But Sentinel, commercially cautious, built only about thirty steam tractors in total, of which only eight were Rhinos.[34] When Leslie brought a prototype Rhino to Pitchill, Tom was much impressed by its versatility. Soon afterwards, Leslie took the Rhino abroad for demonstration to farmers in Kenya and South Africa – an experience which convinced him that if Sentinel were to make a success of their agricultural machinery, they needed to 'accept diesel motors and pneumatic tyres'. When the Company declined to accept his advice on this matter, he resigned his post.[35]

Willans by this time had moved to Stoke-on-Trent locomotive manufacturers Kerr Stuart and Company Ltd to take up the post of 'general manager'.[36] While he was still with Sentinel, keen to improve the efficiency of industrial locomotives, he had begun to explore the idea of 'using one of the new high-speed oil-engines' in place of steam power.[37] Indeed, Sentinels' failure to take much interest in his work was the major reason behind his determination to try his luck with Kerr Stuart – a more enterprising employer – and it was not long before Leslie Bomford joined him. Evidently keen to follow them, on 11 June 1928, Tom took up a three years' premium engineering apprenticeship with the same company.[38] Douglas Bomford supplied him with a strong reference, testifying to his experience of 'fitting, turning, shaping, drilling, etc.', mentioning his work in 'erecting experimental steam ploughing engines and driving same on test'. Not only did he add that Tom was 'Very keen on railway work' but also observed that he 'should make a capable engineer'.[39] In all likelihood, having enjoyed his time at Pitchill, Tom was ready to move into a larger sphere. No longer the self-conscious schoolboy who tried to hide his exhaustion after a morning spent minding the mill and who had to be enjoined to 'Hit it, don't kiss it' at his hammer and chisel tasks, he had developed both a useful stock of practical skills and a certain workplace knowingness. He had also grown familiar with the teasing, foibles and duplicities – the 14T plaque comes to mind – that went on. It would serve him well in the next phase of his career.

Chapter Four

The Apprentice in the Town

With his move to Stoke-on-Trent, Tom entered a brash new world. Experiences new, vital and sometimes profoundly disturbing crowded upon him in the two and a half years that he spent there. When, long after the event, he came to write about this period of his life in *Landscape with Machines*, one of his abiding memories was of the smoke that belched across the sky from the town's bottle ovens. To Stoke's inhabitants, he realised, the smoky skies were a source of pride – 'the outward and visible sign of prosperity' – and the Six Towns celebrated them in the 'smokescape' picture postcards on sale in the corner shops.[1] As an apprentice, he would earn a weekly wage of 10s in his first year; 15s in his second and £1 in his third.[2] For 30s a week – paid by his parents – he boarded throughout his time in Stoke at 439 London Road in Hanley with Mr and Mrs Rock, a pottery carter and his wife. It was, he related, a typical Six Towns house with 'small, cramped rooms', a 'narrow and vertiginously steep' staircase and gas lighting which meant that the Rocks and their tenants needed to keep coins to hand for the meter.[3]

Kerr Stuart operated from the California Works on Stoke's Whieldon Road, the Fenton site bounded on one side by the mainline of the old North Staffordshire Railway – part of the LMS since 1923 – and on the other by the very much older Trent and Mersey Canal. Around the works' offices and antiquated machine shop there were imposing sheds which housed the forge and foundry, boiler shop, wagon shop, erecting shop, paint shop and joinery. A test track ran alongside the railway fence and extended over the entire length of the works.[4] The company was known largely for the engines designed for work in quarries and mines – among them, small Skylarks, Wrens and Tattoos – which it sold ready-made. It also took orders for locomotives to be built to the customer's design in all gauges. The 1920s would see it experiment successfully with Diesel technology, both in locomotives and road vehicles. Tom's apprenticeship would bring him a wide variety of experience.

On his first day, once he had been issued with his 'check', he found himself despatched to the wagon shop where he was put on the task of 'assembling bogies for a series of eight-wheeled wagon underframes [...] destined for an Indian Railway'.[5] It was his introduction to a learning experience in the course of which he would move by stages to the Boiler Shop, Forge, Machine Shop and Erecting Shop. After Pitchill, with its small workforce and modest premises, it was a turnabout – a revelation – and the sheer noise was 'like Bedlam'.[6]

Skill borne of experience counted for much in this world and honing his skills was central to Tom's workplace training. Despite his best endeavours, things could easily go wrong and on occasion, like other apprentices, he found himself having to conceal a useless element, fruit of his bungled labour, under his overalls for clandestine committal to the depths of the Canal.

Writing *Landscape with Machines*, Tom allies himself with his workmates, presenting them very much as a united team. At the same time, it would be surprising if awareness of class difference did not impinge upon his relations with the Kerr Stuart men among whom he worked. Describing them as 'genuine, friendly and kindly', he observes pretty much in the same breath that they disliked and distrusted anything that resembled pretension. To illustrate the point, he gives a sharp sketch of the teasing to which they subjected another apprentice, the son of a local pottery manufacturer, who became the butt of shop-floor ridicule. Greeted by mimicry and catcalls at every turn, Tom suggests that it must have been because the hapless youth's public school education rendered him 'incapable of mixing unselfconsciously with his workmates'.[7] Even allowing for his having spent two years among the Pitchill enginemen, it was not long since Tom himself had been at Cheltenham College which begs the question of how – or, indeed, whether – he escaped the same fate. Claiming that he had already absorbed 'the rudiments of his trade' he goes on to suggest that within a highly skilled workforce, the only way by which a man earns the respect of his colleagues is 'to excel in [his] craft'.[8] It feels somewhat disingenuous. Although he had acquired some useful dexterity with hammer and chisel in the Pitchill workshop, Tom at this time was an apprentice, not a seasoned craftsman. Besides, it is difficult to believe that the traces of his genteel upbringing entirely escaped notice on the shop floor. While he may have got along well with charge-hand Ernie Lines and fitter Jack Hodkinson and found a measure of acceptance among the workforce, being a relative of Development Engineer Kyrle Willans and friend of Works Manager Leslie Bomford may well have smoothed his progress.[9]

Although he does not appear to have worked there, Tom found himself drawn to the works' foundry. Entranced by the precision and finesse that the foundrymen brought to their work, he says that he 'welcomed any excuse to walk through it' so as to see casting in progress.[10] The process fell into a sequence of discrete stages. First, the furnace man released white hot metal into a waiting ladle, capable of holding anything up to five tons of metal. Once full, the overhead crane would swing it into position, ready for pouring. Then, using hand wheels, two men angled the ladle so that it discharged the molten metal directly into the mouth of the mould; a highly skilled operation, it demanded the utmost precision. At this point, a third man would skim off any impurities that formed on the surface of the metal while a fourth man burnt off the gases which escaped from the mould's vent holes. A hazardous procedure, if it went awry, the consequences were likely to be fatal.

It was something which seems to have haunted Tom's imagination. There is no definitive evidence that he actually witnessed the death of a foundry man in his time at the California Works, yet in two of his fictional works, just such an episode becomes central to the action. In 'Strange Vista', the never-to-be-published novel that Tom wrote in the 1930s and based on his time in Stoke, his hero and *alter ego*, David Hancock, has to repair one of the overhead cranes used to position the ladles in the foundry of the Cray or Gehenna Ironworks – Kerr Stuart's fictitious counterpart.[11] Working at roof height with a panoramic view of the routine work in progress on the foundry floor, David observes a 'golden fountain' of molten metal rise from the half-filled mould 'like a giant roman candle' to shower down upon the hapless man who is on skimmer duty. Staggering, the injured man tries to escape, only to collapse screaming before he dies on the sand of the foundry floor. The only explanation offered for the disaster is that the mould has 'gone up' – a rare event, resulting from 'an accumulation of explosive gas which should have escaped through the vents'.[12]

Tom would use the incident again – and indeed presents it in the self-same words that he had used previously – when he came to write 'Hawley Bank Foundry', one of the short stories in the collection *Sleep No More* (1948).[13] Yet nowhere does he mention actually having witnessed such a disaster and there is no evidence to suggest that anything of the kind actually occurred at the California Works. But in March 1929, at the foundry of the Shelton Iron, Steel and Coal Compapny, a 'tapping hole' blew out while casting was in progress and red-hot molten metal and clay showered down upon the men working there. 'Shelton Bar' as the local people called it, mesmerised Tom who used to look across to it from Basford by night, relishing the 'fierce and violent drama' of steel in the making. It is possible, but not certain, that he witnessed the disaster of March 1929 and he may also have read the report of the event that appeared in the *Staffordshire Sentinel*.[14] What is clear is that, real or imagined, the scene somehow lodged itself in his thinking. By rights, such things should not happen, but they did, and the men employed at Kerr Stuart, Shelton Bar and similar firms encountered danger on a daily basis. Tom's only comment, terse and irrefutable, was that 'Molten metal is deadly stuff'.[15]

His time in the erecting shop would be one of the high points of his apprenticeship since it coincided with the completion of work on a prestigious contract for six super-heated 4-8-0 mixed traffic locomotives destined for the

Buenos Aires Central Railway – the *Ferrocaril Central de Buenos Aires*.[16] To his satisfaction, Tom found himself on a gang responsible 'for building a locomotive from the time the bare frame plates were set up until it left the shop under its own steam'.[17] Working under the supervision of charge-hand Ernie Lines, he marked the steady growth of 'the monster' until it was ready for the works cranes to lift it, and set it upon its wheels. Once they had fitted the valve gear and connecting rods, fussing, in Tom's memory, 'like anxious midwives', the men raised steam and allowed the engine 'to run herself in … driving wheels revolving slowly' not yet on the test track, but on rollers set at track level just outside the shop entrance. Once Mr Lines was satisfied, his gangers fitted the coupling rods ready to run the engine under steam while Tom took the controls. It was his first experience of driving a locomotive and the sensation of feeling the 'great machine suddenly come alive under [his] hand' left him in high ecstasy.[18]

While the bulk of Kerr Stuarts' orders were for steam locomotives, by this time the firm was also committed to promoting diesel technology and in July 1928, Tom attended the test trials of Willans' prototype diesel engine – KS 4415 – on the Welsh Highland Railway.[19] With its steep gradients and tight curves, the Bryngwyn branch afforded every opportunity for a demonstration of its potential. Mindful of its efficiency in terms of fuel consumption – specifically, it promised to save 15s in fuel costs for every £1 spent on coal for steam locomotives – Kerr Stuarts' sales manager, G.R. Greenbergh, promoted the diesel with the auspicious slogan, 'fifteen shillings change'. 15 November 1928 saw a carefully orchestrated demonstration of KS 4415's capabilities for the benefit of the press. If the *Staffordshire Sentinel*'s praise for 'the ease with which it was shunted and reversed' owed something to its partisan support of a local enterprise, admiring descriptions of the engine appeared also in such specialist periodicals as *Oil News*, *Machinery Market* and the *Quarryman's Journal*.[20]

In sales terms, the publicity paid off generously. The Sudan Plantations Syndicate, for instance, purchased a number of 60 hp Kerr Stuart diesels – sister locomotives of KS 4415.[21] Tom recalled delivering and setting to work a 30 hp model at a quarry near Nuneaton. Another, he took for trial at the site where the East Lancs Road was under construction, but it proved too heavy to be useful. One 90 hp model went on to have a long working life on the Ravenglass and Eskdale Railway; a second went initially to the Air Ministry at Cranfield before moving to a brickworks.[22]

Significantly, in the 1929 Stoke-on-Trent Association of Engineers' annual essay competition, Kerr Stuart apprentice Cecil Howell won first prize for 'The Diesel Locomotive' while Tom was runner up with 'The Light Railway of Today' in which he made a powerful case for adoption of diesel power on narrow gauge lines.[23] Specifically, he argues that the maintenance of narrow gauge steam locomotives – 'even if kept in good working order, which is rarely the case where the staff is unskilled and there is little or no adequate shed equipment' – posed particular challenges to small independent railways. What was more, they were uneconomic to run and maintain. 'For such intermittent use as the light railway demands' he observes, probably echoing Willans' view, 'the amount of coal and water consumed while standing in steam ... makes this type of motive power an expensive proposition'. For efficiency and economy in conditions where 'intermittent service' was unavoidable, there was 'only one real solution', namely the Diesel Locomotive.[24]

Imaginatively, towards the end of his essay Tom suggests that the nation's light railways could look to the tourist market for fresh custom. Narrow gauge lines, he argues, often hold unique fascination to the public. 'The type of country where the advantages of the Light Railway are most apparent', he argues, tends to be mountainous and sparsely populated – exactly the sort of place 'that holiday makers would flock to see'. Although a number of light railways traversed 'some of the grandest scenery in Wales', among the public at large they were almost unknown. Some 'judicious advertising ... at holiday periods' he argues, and the 'co-operation with charabanc owners to run combined road and rail trips' could give rise to a new type of holiday attraction. Advancing this idea, he may have had the GWR's 1924 guidebook *Welsh Mountain Railways* in his thinking.[25] Significantly, he viewed the usage of diesel power as integral to his hypothetical venture's success. 'The employment of internal combustion engined locomotives', he asserts, 'and the consequent absence of smoke and smuts would make possible the use, in summer, of open coaches on road vehicle lines'. Mindful of the Welsh weather, he also suggests that fitting rain-proof hoods would be a wise precaution. How far Tom's youthful conception of rail tourism contributed to the zeal with which, years later, he would devote himself to the Talyllyn venture is hard to say but his essay at least carries some recognition of the affection that railways, particularly those which ran through attractive country, held in the minds of the public at large.

Tom found weekends in Stoke tedious. When the industry which lent the place its vital energy turned quiet, the smoky skies cleared and 'the pitiless eye of the sun revealed with unfamiliar clarity the sordidness of the cobbled streets with their rows of terraced houses', he 'could not wait to get out'.[26] His cousin Bill Willans was, at the time, an apprentice with the Shropshire and Montgomeryshire Light Railway, which ran between Shrewsbury and Llanymynech just across the Welsh Border, with a branch to the village of Criggion, and on Saturday afternoons Tom would head off towards Shropshire on his motorcycle. Together, Bill and he examined the 'ancient' locomotives in the shed, appraised the repair shop with its 'primitive' equipment and often borrowed a platelayers' trolley on which to explore the line.[27] Years later, Tom remembered crossing the timber trestle bridge over the Severn at

Melverley – an experience which left him with mixed feelings. In *Red For Danger* he would state soberly that its decay caused him genuine 'trepidation', while in *Landscape with Machines*, he describes traversing the old bridge as a 'thrilling highlight of the journey'.[28]

Besides these railway musings, Tom's weekend visits to Dovaston House, the Willans family's Shropshire home, saw the first stirring of his interest in canals. In the days of his partnership with Mark Robinson at the Ferry Works, Thames Ditton, Peter Willans – Kyrle Willans' brilliant short-lived father – had owned a steam launch, *Black Angel*. Aboard *Black Angel* he made an adventurous voyage in the 1880s, following the waterways north to Ripon and back.[29] Kyrle, who had inherited all his father's canal zeal, purchased a Shropshire Union Fly Boat – *Cressy* – from the corn-milling Peate family of Maesbury at the end of the 1920s; arranged for boat builder John Beech to convert her 'into a comfortable houseboat' at his Frankton works and installed a compact steam engine to power her.[30] Tom took part in her steam trial, 'up the canal in the direction of Llangollen' on a Sunday afternoon in 1929 and yearned to steam across the Pontcysyllte Aqueduct which, some eleven tantalising miles away, was 'manifestly impossible' to reach in the short time available when travelling at 3 mph.[31]

The following spring, the Willans family left Dovaston House to settle in Barlaston on Stoke's rural fringes. It meant the loss of a beautiful family home in enchanting country for which, Tom suspected, 'the Potteries must have seemed ... a very poor exchange', but it spared Kyrle the business of 'trundling off each weekend in his Model T Ford sedan to re-join his family', while lodging from Mondays to Fridays in Trentham.[32] The Willans family had also to move *Cressy* to the Trentham mooring that was to be her new home and Tom, Bill and 'a youth named Frank' assisted Kyrle in steaming the vessel on the landmark journey – her 'maiden voyage as a powered craft'.[33]

Their route took them across north Shropshire by Colemere, Blakemere and Whixall Moss, to Middlewich where they joined the Trent and Mersey Canal.[34] In his recollection, Tom came to associate the journey with the dawning of a new and potent insight. Stoke-on-Trent with its works, potteries, mines and Shelton Bar had shown him the England that the Industrial Revolution had created and the energy of its 'spinning wheels ... smoke and steam and furnace flame' thrilled him to the core. At the same time, he could not help but mark the difference between the hectic world of industry and the quiet hills of his childhood. The idea that the one might, in time, overreach the other struck him as 'too terrible to contemplate'.[35] Yet, as he remembered it, standing upon *Cressy*'s deck on that March morning in 1930 and looking out at the surrounding mere, the vexatious tension between the opposed elements – town bustle and country quiet; manufacturing industry and pastoral subsistence; old beliefs and modern values – miraculously fell away. *Cressy*, he reflected, 'did not intrude upon landscape; she became a part of it, like the canal itself' as it crossed remote Whixall Moss under a 'colourless sky'. It suggested that his 'consuming interest in engineering' on the one hand and his 'feeling for the natural world' on the other need not be at odds with one another. To Tom, it was a revelation and he soon realised that he had 'fallen head-over heels in love with canals'.[36]

The next day, things turned somewhat farcical. From Wardle Lock in Middlewich where the party had moored overnight, they came without difficulty up through the locks by which the canal climbed to the summit level at Kidsgrove only to face a significant challenge at the northern end of the Harecastle Tunnel. To transit the tunnel, an unstoppable electric tug drew boats along its 2926 yard length in a waterborne convoy which could consist of anything up to thirty vessels. Headroom in the tunnel was extremely limited and if the cabin which with *Cressy* had been provided in her conversion from working boat to houseboat was not to be ripped off, her crew realised that they would have to ballast her down so that she rode very low in the water. It put paid to any hope of their reaching Barlaston that evening. While Tom's work commitments meant that he had to jump ship without delay, the rest of the party spent the night at Kidsgrove and the following morning filled *Cressy*'s hold with bricks. It proved sound strategy, for they came through the tunnel unscathed and, having ditched their weighty cargo, went safely on to their destination.[37]

In spring 1930, Britain stood on the verge of economic depression but Tom, well on the way to completing his apprenticeship and confident that he would soon be readily employable as a skilled man, was in buoyant spirit. Already, he was entrusted with such tasks as road-testing Willans' new diesel lorries and demonstrating them at agricultural shows, but the diesel locomotives were a notable success. Meanwhile, the GWR had recently ordered some 'twenty-five new six-coupled pannier tank locomotives'.[38] To all appearance, Kerr Stuart's outlook was bright.

But the company was not as secure as the omens might seem to suggest. Addressing the shareholders at the 1929 AGM, Chairman and Managing Director Herbert Langham Reed had offered a less than optimistic forecast of the firm's fortunes. Not only had the previous year proved less than profitable, but a fall in the market meant that the prices that Kerr Stuart obtained for their locomotives were low and 'showed no tendency to rise'.[39] Work, Reed conceded, was readily available, but that was not to say that it was profitable. Besides, there were a number of discouraging factors in play. A local coal dispute had caused a dearth of raw materials which nearly caused the works to shut down.

A large debt was owing in relation to the funds which the Company had laid out in the Peninsular Locomotive Company of India although negotiations for compensation were, apparently, 'pending'. Reed could identify only two genuinely encouraging signs for the future. Firstly, Kerr Stuart had 'an interest' in Low Temperature Carbonisation Ltd., – a forerunner of the Coalite Company – whose Barnsley plant was apparently producing an impressive-sounding 1000 tons of Coalite a week. Secondly, the firm had invested in Evans, Osgood & Co Ltd., a builders' supply company whose business, Reed reported, 'had so developed that it had been decided to form a new company to handle it'. Taking these factors into consideration, he was cautiously confident that Kerr Stuart would 'rapidly return to a dividend paying stage'.[40]

Evans, Osgood & Co Ltd, the 'new company' on which Reed pinned his hopes, also traded as Evos Doorways Ltd.[41] Quite what factors led him to suppose that sinking a substantial part of Kerr Stuart's capital in a firm that manufactured doors was a good policy, it is impossible to deduce. Having made a poor showing in its first trading period, in December 1929 Evos Doorways Ltd. reported a loss of £7151.[42] In March 1930, it definitively failed, prompting the Midland Bank which looked after its finances, to apply to Kerr Stuart for the funds that they had lost in the Doorways' company's collapse.[43]

As Tom presents the story in *Landscape with Machines*, the first that he knew about the Midland Bank's petition came from seeing a notice on the *Daily Mail*'s finance pages. At the first opportunity he drove out to Barlaston to show the relevant paragraph to Willans whose response was a mix of incredulous dismissal – 'no-one has heard of anything' – and robust assurance that the Midland Bank had nothing to do with Kerr Stuart's finances.[44] But on 24 March 1930 the *Staffordshire Sentinel* reported that the company's net loss amounted to £15,906 – a figure large enough to give much concern locally.[45] At pains to emphasise how disastrous the loss of the works would be for the men of Stoke, the 3rd Stoke-on-Trent Branch of the Amalgamated Engineering Union petitioned Jimmy Thomas MP to bring his influence to bear upon staying the compulsory winding up order. In the short term, they appeared to have got their way. All but one of Kerr Stuart's creditors opposed the move to close the business and Willans, fully aware by now of the straits in which the Company stood, was instrumental in persuading the Bank to withdraw their petition.[46] By now, it appeared that Kerr Stuart 'had in hand more work' than at any time over the previous eighteen months.[47]

But three days after the jubilant announcement came news that the National Provincial Bank, who transacted the company's financial business, had appointed Sir Harry Peat as receiver and manager, pending what the *Sentinel*, perhaps euphemistically, termed a 'reconstruction'.[48] Together with Sir Francis Joseph, the president of the North Staffordshire Chamber of Commerce, Willans planned a reorganisation of the works in the hope that it might herald a return to prosperity but by 2 June, many Kerr Stuart men had been laid off, and it seemed likely that the workforce would come in for further reduction over coming weeks.[49] Three weeks later, on 21 June, the freehold and goodwill of the California Works were offered for sale by tender as a single lot, but no buyers came forward.[50] Nevertheless, hope lingered on and in July 1930, Willans' Diesel Lorry caught enough admiring attention for the *Sentinel* to suggest that it might not be necessary to close the works 'even temporarily'.[51] At about the same time, Tom transported a 30 hp diesel locomotive by diesel lorry to Scarborough to show it off at an exhibition where an eager purchaser bought it on the spot. It meant that he had to return to Stoke by way of King's Lynn – a vast detour – delivering it to its new East Anglian owners on the way.[52] But by the next month, only a skeleton staff remained on the premises to complete the company's outstanding orders.

Then on 14 August 1930, under a 'Stop Press' headline, the *Sentinel* announced that the London firm of George Cohen & Co., machine tool manufacturers and scrap metal dealers, proposed to purchase the California Works for making locomotives' spare parts.[53] If it was hardly a return to Kerr Stuart's glory days – the newspaper acknowledged that it was uncertain whether 'the whole or any considerable part' of the site would re-open – it must nevertheless have generated considerable excitement.[54] But in autumn 1930 Cohens decided that they no longer had any use for the works. Soon, another, bleaker 'Stop Press' notice in the *Sentinel* disclosed that a winding up order had been issued against Kerr Stuart.[55] In November, the firm's creditors met the Official Receiver. Liabilities to unsecured creditors amounted to rather more than £85,500 and contingent liabilities to £146,142; unsurprisingly, the Official Receiver called it a 'complete shipwreck' and it was agreed that the company's effects should be auctioned off.[56] In the event, the land and buildings were withdrawn from the sale, while plant, machinery and stores were purchased piecemeal.[57] For a once innovative and highly successful company, it was a melancholy end.

Willans wrote to the *Sentinel* in high indignation deploring the fact that throughout negotiations concerning Kerr Stuart's future, the men who worked there and who were responsible for the firm's past commercial success, had had no 'say' whatsoever regarding their fate. No thought, he protested, had been given either to the deepening economic depression - Kerr Stuart had been a large employer in Stoke - or to 'the fact that the firm had orders; or to the happy conditions existing between the men and the Management, which were in themselves a guarantee that work could be got and be made remunerative, even at the low prices obtaining'.[58] Although he made a powerful case for keeping the company in production, it was too late to save it.

Tom's sense of shock and betrayal at the firm's demise was acute. which may explain his inclusion of some extravagant details in the *Landscape with*

Machines' record of events which have questionable factual basis. He gives, for instance, a memorable vignette of Kyrle Willans calling on Lady Cynthia Mosley, Stoke's Labour MP of the time, and begging her to solicit government aid to keep the company afloat. According to Tom, who was not a witness to the conversation, she reclined throughout their exchange upon a sofa, smoking and contributing only 'soothing platitudes' to the discussion. Apparently, they entirely failed to soothe Willans who returned home more 'angry and ... bitter' than Tom had ever seen him.[59] Frustratingly, although it is understandable, no other account of the interview survives to corroborate this version of events, but the following year, Lady Cynthia lost her seat to the Conservative Ida Copeland. There is also a total lack of evidence to support Tom's allegation that around the time of Evos Doorways Ltd's demise, Reed's private secretary was found dead in his office with 'a bullet through his brain and the hearth choked with burnt papers'.[60] It is, nevertheless, a matter of public record that late in November 1930 Cecil Langham Reed, Herbert's brother, shot himself in the garden of his Hampshire home, driven – apparently – to suicide by anxieties about his business. Yet the coroner who conducted the inquest into Reed's death dismissed his worries as 'needless'.[61] Five years later, plagued by money troubles of her own, Cecil's widow Gwendoline Rose ended her life by jumping off a cross-channel ferry.[62]

As for Herbert Langham Reed, remembering his occasional visits to the California Works by Rolls Royce from which he emerged in his pin-striped suit to make tours of inspection 'looking neither to right nor left', Tom viewed him as an unmitigated villain who, he claims, disappeared without trace after Kerr Stuart's failure never to be heard of again.[63] History does not entirely support this interpretation of events. Besides his role of Chairman and Managing Director of Kerr Stuart & Co Ltd., Herbert Langham Reed was also a director of Arsenal Football Club. At the time of his death in January 1934, it was widely reported that he had been dressing to attend the funeral of Arsenal's distinguished team manager Herbert Chapman.[64]

The Hunslet Engine Company of Leeds took over the goodwill of the business and also accumulated all Kerr Stuart's 'manufacturing rights, patterns, drawings and press blocks'. With the intellectual property and artefacts, they acquired pretty much the entire understanding of diesel technology as it existed at the time.[65] Although Tom could not have foreseen it, by picking up the pieces after Kerr Stuart's ignominious end, Hunslet equipped themselves to do him some significant future favours.[66] But at the time he could find no consolation for the wreckage of the company on which, as an apprentice, he had pinned his fondest hopes.

Writing his autobiography years after the events that it describes, Tom is lucid about the shock that he felt at seeing a productive company crumble away and his raw anger over the treatment of the men with whom he had once worked blazes bright on the page. He was also, with good reason, aggrieved on his own account. When first he came to Stoke-on-Trent, the future – his future - had brimmed with promise. Seeing Willans' development of diesel locomotives and lorries at first hand thrilled him to the core. To forfeit the opportunities which should have followed completion of his 'time' at Kerr Stuart dealt his ambition a mighty blow. At the same time, Tom recognised his good fortune in being Willans' protégé and Leslie Bomford's friend and does not dwell on his own missed opportunities which were small compared with the plight of the men whom Kerr Stuart's collapse threw out of work. Bomford, through his friendship with the Lister family who managed the Gloucestershire agricultural-engineering firm of R.A. Lister & Co. Ltd. arranged for Tom to complete his 'time' at their Dursley works. If he lost the £100 premium that his parents had paid at the start of his apprenticeship and promised to give him when he finished it, he still had the chance to gain the status of an engineer. Nevertheless, the experience of seeing so proud a firm as Kerr Stuart founder left him lastingly shocked and sore. Gloomily, he went off to Gloucestershire, the sense of betrayal heavy on his pulses, and a strong suspicion that he was entering an alien, inimical world.

Chapter Five

Marking Time

Tom arrived in Dursley in January 1931. The contrast between the leafy market town and industrial Stoke could hardly have been greater. Where Stoke had potbanks and ironworks amid the railways and canals, Dursley's historic industry lay in wool production which had centred in mills on the river Cam. In place of Stoke's vaunted smokescapes, 'smouldering waste tips, pit mounds and heaps of furnace slag' alongside 'endless cobbled streets lined by terrace houses of soot-blackened brick,' Dursley boasted the 'Garden Suburb', an estate of fifty-eight houses which Sir Ashton Lister financed in 1911 to accommodate his workforce.[1] It may have owed its inspiration to Margarethenhöhe, the village which Margaret Krupp financed and built in 1906 as social housing for employees of the Krupp steel works. Suggestively, a 1931 newspaper article remarked that 'The Towers' – originally Sir Ashton Lister's mansion which his heirs had converted into a hostel for Lister apprentices and a lodging for Lister clients – invited 'comparison with the Essener Hof', that is, the hotel 'attached to the Krupp works in Essen'.[2] Viewed retrospectively, the enthusiasm of the piece strikes rather a chilling note. Tom does not mention whether he ever read it.

So conspicuous is the current of frustration which runs through his memories of working for R.A. Lister & Co. Ltd of Dursley in *Landscape with Machines* that it raises the question of whether he was not prejudiced against the company from the outset. Assuredly the work that Tom did in Dursley brought him none of the brimming excitement that he had known in Stoke. Recognising that he had considerable practical experience, his managers took the pragmatic view that they could entrust him with skilled work. They therefore assigned him the tasks both of building up large diesel engines which were too complex for assembly line production and of testing the newly assembled engines' performance by running them on the dynamometer and correcting any faults. Neither job was entirely lacking in interest, but – he reflects, aggrieved – both of them - even if they constituted 'skilled work' in Listers' terms - represented a 'blind alley' as far as his future engineering career was concerned.[3] Besides, the company paid him only an apprentice's wage. At Kerr Stuart, Tom had been one of a team committed to 'exciting pioneer work' and rashly he assumed that Leslie Bomford would have told Robert Lister – head of the firm's trucks division – that he had assisted in the development of the Kerr Stuart diesel lorry. He even cherished the hope that he might have the opportunity to pursue a similar project at the Dursley works. While he was there, Listers did indeed build a prototype diesel lorry engine, but Tom never had any chance to involve himself with it and before long the company abandoned the entire project.[4]

Much of his antipathy to Listers stemmed from their production method. Shortly before he started working for them, they had introduced an element of mass-production to their works. Rather than rely on skilled craftsmen of long experience and great individual aptitude, for the manufacture of their standard rages of small petrol and diesel engines they recruited and employed a semi-skilled workforce on an assembly line. It meant breaking down the manufacture of a diesel engine, for instance, into a series of defined steps or processes, none of which demanded any high level of proficiency, for completion, in a pre-defined sequence.[5] To Tom, it was an anathema – an evil which, in his view, promised to change the whole nature of manufacture manifestly for the worse. Although he conceded that rural Gloucestershire had little of Staffordshire's tradition of engineering, he remembered how locomotive building at Kerr Stuart 'had been a craft industry' in which the 'one true entitlement to superiority' was 'to excel in one's craft'.[6] What he found at Listers' Dursley works left him smarting with indignation. In the context of assembly line production, skill was pretty much irrelevant; it no longer served any practical purpose. In place of skilled craftsmen, the Listers' assembly line relied upon part-trained operatives. Instead of taking a justifiable pride in well-honed dexterity which was no longer required, they were, he claims acutely 'class-conscious' and 'inclined to judge their fellow-men by appearance and material possessions'.[7]

As for the various initiatives that the Lister family introduced and provided for their workforce – provision of in-house dental and medical services; a Social Club which sold subsidised beer, not to mention the excursions, the company cricket team and the *Lister Standard* magazine – he does not mention them.[8] Perhaps he felt that by tacitly extending its sway over the leisure of its employees as well as their working hours, company paternalism over-reached itself. Although the Duke of York – later King George VI – visited Dursley and the Lister works in December 1931, Tom says nothing about it in *Landscape with Machines*. If His Royal Highness 'mingled' with the employees over lunch, played skittles in the social club bar and graciously allowed himself to be introduced to Frank Harris, the oldest man on the company books, Tom was not interested. Neither did he have any time for the charade of HRH's test-driving an auto-truck, casting a plate in the foundry and stopping and starting the machines on his ceremonious walk through the works.[9]

Throughout his time in Dursley Tom, like other Lister apprentices, lived at 'The Towers' which served as a works' hostel, albeit on an unusually grand scale. He appreciated both the companionship and the comfort that the accommodation provided and recalls that for the first time in his life, he found himself 'basking in the reassuring warmth that comes from friendship with a group of people of my own age'.[10] Since a number of them were sons of Listers' overseas agents on whom the directors wished to make the best possible impression, the company employed one Mrs Vanderguilt to oversee the Towers' smooth running in the role of hostess-cum-house-keeper.[11] Tom misremembers her surname as 'Van der Gucht' which appears to have been the name of a family who lived at Kemerton near Tewkesbury. Describing the Towers' housekeeper 'charming' and 'vivacious', he clearly liked her, but his goodwill had its limits. With her endeavours to introduce the young men of R.A. Lister & Co. 'to polite society' – in other words, well-to-do local families with daughters – he had no patience whatever. Gloucestershire mamas might view The Towers as a 'reservoir for eligible young men', but Tom and his Sydnesider friend Ron Malloch side-stepped their blandishments and declined their proffered invitations, preferring to drink in the dockside pubs of Gloucester and Bristol.[12]

What cemented his friendships with his fellow apprentices was a shared interest in cars which, Tom states, 'nearly all' the Towers' residents owned.[13] In Stoke, the fact that he owned a G.N. – one of the range of modest cyclecars that H.R. Godfrey and Archibald Frazer-Nash produced in the 1910s and 1920s – had seemed anomalous. Since Kerr Stuart employees, to a man, either walked or cycled to work, on the rare occasions when Tom arrived in his car, he found it rather embarrassing.[14] But in well-heeled Gloucestershire, he had no cause for such self-consciousness and the G.N., which he had purchased in Hanley for the sum of £6, found its place in the Towers' forecourt among a diverse medley of other vehicles. While living in Dursley, he purchased a second G.N. which – surprisingly in view of his disinclination to view the local constabulary in any friendly light – came from one of the local policemen. He paid £8 for it, which at the time struck him as a hard bargain, but it was in manifestly better condition than its predecessor which he came to regard as a source of spare parts.[15] Other cars on the Towers' forecourt at this time customarily included a sporty Riley Redwing, property of a Lister pupil named McClure; a six-cylinder AC – Auto Carrier – belonging to Malloch; Mrs Vanderguilt's stylish Standard Little Nine open two seater; a larger Standard Fourteen, whose owner Tom does identity although he states that the vehicle came from a Gloucester scrap yard; a Ruston Hornsby which Tom, again declining to name the owner, pronounced 'slow, cumbersome and lorry-like'; and Timothy Huish Edye's long-bodied Belsize.[16]

For the Towers' residents, driving was nothing if not eventful. Edye crashed the Belsize on a foggy December night skidding on ice near Nympsfield.

Marking Time

He had been driving his girlfriend to a dance in Cheltenham at the time and, Tom recounts, 'got confused by the mist [and] lost his sense of direction'.[17] Malloch replaced the AC with a 14/40 Vauxhall after an ugly accident. Approaching a sharp right hand bend, its steering had locked. At the time, Tom had been driving – Malloch had a suspected hernia – and found himself 'tugging fruitlessly at a completely immovable steering wheel'. The car hit a bank, shot across the road and overturned, depositing driver and passenger in a ditch. It was a write-off, although Tom, who always liked to identify the causes of aberrant behaviour in machinery, traced the trouble to the car's worm-and-wheel steering system. 'The teeth on the used arc of the worm-wheel' he observes in a precise, technical note, 'can (as had happened in our case) become so worn that the worm rides up on them and becomes locked.' It was an aberration well-known to seasoned AC owners who, if the steering came to feel 'lumpy', took the steering box apart so as to reposition the worm-wheel so that 'the arc of unworn teeth […] came into play'.[18]

By January 1932, between his time both in Stoke and at Dursley, Tom had completed three years' work in different branches of mechanical engineering. It meant that he was now, officially, a skilled man but of this significant milestone, Listers made no acknowledgement. The silence left him understandably aggrieved. He had served his 'time' and the company engaged him in skilled work, but paid him at the rate of an apprentice. Smarting with the injustice, he steeled himself to approach the management.

It was not a propitious encounter. Perhaps, as the country slid ever deeper into economic blight, it was naïve of Tom even to imagine that Listers would offer him a well-paid position. When he asked about the chance of obtaining a permanent post, Listers not surprisingly responded that they were sorry, but they could not offer him a job.

Despite his frustration with the firm's complacency and petty snobberies, the discovery that they had no further use for him came as an ugly shock. Unimpressed, Tom made his farewells and packed his belongings into the G.N. Before he drove away to Stanley Pontlarge, he exacted some small-scale revenge by abandoning a derelict car on Listers' premises and appropriating the metallic fig leaf which had preserved the modesty of a statue of an otherwise naked gladiator in the Towers' entrance hall.[19]

Over the next two years, Tom had a number of short-lived jobs. 'Whenever I was lucky enough to obtain one', he observed, 'it did not last long.'[20] Between April 1932 and August 1933 he worked for two small agricultural engineering firms, first for I.A. Bennett of Hungerford, who were based at what was known as the 'Tractor Store' and later for the 'Aldbourne Engineering Company'

who were based at the Foundry in the Wiltshire village of Aldbourne. In both positions he found the work absorbing and came to relish the high chalk downs. He also enjoyed the company of Mark Palmer, an elderly employee of the Aldbourne concern much given to reminiscence about his employment with traction engine manufacturers Wallis and Steevens. Tom always valued the conversation of aged countrymen with a habit of storytelling, as Bill Smith of Pitchill and, in later years, Harry Rogers the Ironbridge coracle maker could testify. But his Aldbourne job came to an end in August 1933 and between October and December 1933 he found employment fitting up boilers for steam wagons at the Sentinel Waggon Works, Shrewsbury. A falling off of demand soon meant that he was out of a job once more, but the following year he had a brief period of employment with Thornycrofts in Basingstoke which he found entirely 'unmemorable' and which ended in April 1934. Then for a short time, he worked as a tractor salesman. Otherwise, it was a period of his life that he came to view as 'roving, rootless and largely fruitless'.[21]

By way of diversion, he purchased a 1903 Humber, once the property of a Winchcombe market gardener named Greening, which he found in a chicken run; restored to working order and entered it in successive London – Brighton veteran car runs.[22] He also read widely. He loved poetry and developed a special fondness for the works of W.B. Yeats. Keen to understand what had brought the Industrial Revolution into being, he studied Samuel Smiles' three volume *Lives of the Engineers* which he bought second hand for 7s 6d. He read Cobbett, whose antipathy to enclosures and the 'Speenhamland System' of poor relief he shared, and he devoured Aldous Huxley's *Brave New World* (1932) which he considered 'quite brilliant'.[23]

Reading coincided with writing. The sore and confused feelings with which his recent work for Kerr Stuart and Listers had left him led Tom to construct a long and ambitious narrative with a strong autobiographical slant. At the time, he had recently come across a poem entitled 'The Song of the Wheel', a verse-meditation upon the wheel's role as a power source throughout history, which he liked enough to include in his commonplace book.[24] In the opening stanza, the wheel which serves as its narrator, asks rhetorically:

> Through what strange vista did he see
> > Of days unfurled,
> Who first of all men fashioned Me
> > That shook the world?[25]

The lines gave Tom his novel's title – 'Strange Vista' – and within it, he sets out to explore humanity's relationship with industry while at the same time, tracing successive generations of two families – the rich Crays and the poor Hancocks. He wrote it in longhand and two discrete manuscripts survive in the

Rolt family's possession. One, which is complete and probably represents the earliest version of the tale, occupies a 'Heraldic Series' board bound notebook, its covers decorated in red. The other, apparently a revised version of part of the text, is in a similar notebook but with green cover decorations. Neither manuscript supplies the date of its composition.

From the complete version, it is evident that Tom intended that his story should fall into three parts, set respectively in the past – the 1760s; the present – that is the late 1920s – and the future as suggested by the 1950s. In the 1760s section, the 'Prologue', Tom shows the squirearchy, in the person of John Cray, welcoming the opportunity of enrichment that the canal promises to bring. The labourers of the district are more suspicious. 'T'inna natural', pronounces old Mr Hancock of Lea Cottage when he learns about the waterway. 'God did'na intend them boats to go a-floating about this country o'ours that way. [...] Mark my words, no good'll come of it.' In the 1920s part of the book, Tom shows Mr Hancock's descendant David as the hapless witness to a fatal foundry accident. It leaves him uncomfortably aware of man's vulnerability in the face of the very forces that he presumes to harness. David falls in love with John Cray's descendant, Anne, only to perish underground when he goes to the rescue of his collier brother John after an explosion in Cray Deep Pit. At the close of this part of the story, Anne dies giving birth to David's son. By the 1950s, Wolvercroft's steam-driven industries have disappeared, and in their place stands a generating plant – the New Cray Powerhouse – which transmits 'clean, swift and super-efficient' electricity to its homes, its factories, its entertainment venues and indeed its vehicles. Since manufacture by this time is largely automated, only a minute number of people have gainful employment while, according to magnate John Cray, most of the population are 'indolent, pleasure-seeking and sensuous'. News comes of a mysterious and fatal pandemic which has broken out in Leningrad. Unknown to medical science, Cray's friend Dr Fordyce describes it as an 'acute form of Brain Fever caused in some obscure way by the conditions of living in the great industrial centres'. It spreads inexorably across Europe, causing devastation in its wake, and wipes out most of the inhabitants of Wolvercroft. Among the few survivors are John Cray's daughter Mary and David and Anne's son, Dick, who agree upon the need 'to build a new world that will be a simpler, more peaceful place' than the machine-driven culture of the past. The story ends with their falling in love, united in their resolve to abandon the 'scientific developments' of previous generations and to support themselves from the land.[26]

'Strange Vista' does not easily lend itself to summary and this brief overview does it less than justice. Admittedly the more melodramatic elements of Tom's storytelling and the assumptions that he makes about town dwellers and country people feel rather jejune and simplistic; his plotting can be clumsy and his characterisation, particularly of the women, lacks subtlety. Nevertheless, his prose is vigorous and his descriptions of life in Wolvercroft – particularly of David Hancock's experiences at the Cray Ironworks – hit home with great power. That he found the stamina to finish writing the novel is greatly to his credit. Remembering its composition in *Landscape with Machines*, Tom remarks in his own defence that over the years of the 1930s depression, it did indeed seem to him 'that the industrial world was changing and might be sowing the seeds of its own dissolution'.[27] Setting aside the questions of whether the fictional 'Strange Vista' carries any significant prophetic charge, the act of putting the tale together was clearly something that Tom enjoyed. Writing carried its own rewards. It gave Tom a sense of purpose and direction in what, otherwise, was a fallow and unrewarding time in his life. With encouragement, he would come realise that he wrote well enough to make a career from it.

In the course of his short-lived job with Thornycrofts in Basingstoke, Tom struck up a friendship with a young man named John Passini. Not only did they share lodgings in the Wortley Road but they soon realised that they both had an interest in motor racing. One Sunday evening, Tom recalls, he returned from a visit to his parents to find John 'full of a marvellous pub he had discovered at Phoenix Green near Hartley Wintney'. The Phoenix Inn, as it was called, owed much of its charm to its landlord, Irishman Tim Carson, who combined managing the pub with setting speed records at Brooklands. A visit soon followed and Carson, an affable man, was quick to welcome Rolt and Passini into his circle-cum-support-team, the self-styled *Scuderia Carsoni*. Their workshop, or 'racing shed' was in a 'humble wooden hut' at the back of the Phoenix Inn where Carson built his 'special racing 30/98 Vauxhall'.[28] While he admired the car, Tom clearly did not think much of Carson's garaging facilities.

But nearby, there was a petrol filling station. Consisting of 'an ugly corrugated iron building with four manual petrol pumps on its forecourt' its manager was a Mr Baldwin, a former county cricket player and MCC umpire, who ran his business in what Tom describes as 'a somewhat desultory fashion'. Believing that they could make more success of the site than he, at some point in 1934, Tom and Passini decided that they would go into partnership to purchase it. They 'put up the capital in equal shares' – for Tom, it meant appealing to his parents for funding – and acquired not only Mr Baldwin's premises, but also an old barn, some office accommodation, a paint shop and a cottage. The purchase greatly depleted their resources with the consequence that they would be seriously short of working capital for the future. Nevertheless, the prospect of launching the new business was exhilarating; besides selling petrol, they planned to run a repair shop and attend breakdowns. It would be a dynamic successor to Mr Baldwin's sleepy concern.[29]

The Vintage Sports Car Club – VSCC – came into being a few months before the Phoenix Green Garage opened under Passini's and his management and would, for Tom, be a source of solid friendship. In October 1934 a letter headed, tellingly, 'A Club for the Not-so-Rich' had appeared in *The Light Car* advocating the formation of a club for the owners of early sports cars – specifically, cars built before 1931, and not, therefore, state-of-the-art racing models.[30] Tom claims that he, Passini and Carson invited its writers, Colin Nicholson and Ned Lewis, to the Phoenix Inn to discuss the idea and that the VSCC established itself soon afterwards. While he may well be right, there does not appear to have been any element of cause and effect at work. The VSCC committee held its first meeting on 23 October 1934; Tom would be elected to club membership on 21 February 1935.[31] Initially known as the 'Veteran Sports Car Club', members of the 'Veteran Car Club' (VCC) – for owners of cars built before 31 December 1904 – were concerned about the risk of possible confusion between the two organisations. Obligingly, the VSCC adjusted its name to 'Vintage Sports Car Club' which usefully preserved the initials while causing minimal trouble.[32]

Tom, at this time, still drove and raced the GN that he had acquired from the Dursley policeman, but it came to grief when a Jowett collided with it near Winchfield.[33] He had a workaday 1922 Belsize-Bradshaw for which he had paid £2 during one of his periods of unemployment and which he considered 'a nice little car for its date' but lacking in anything resembling 'sporting character' it was not, in his view, 'a suitable mount for a [...] member of the VSCC.'[34] With Carson's help, he found a 1924 'duck's back' 12/50 Alvis two seater for which he paid £10.[35] It would be one of the great and enduring loves of this life.

In 1936 Tom would support Tim Carson's motion that the VSCC confined its driving membership to 'owners of cars manufactured prior to 31 December 1930'.[36] The date was not by any means arbitrary. Previously the criterion for driving membership of the VSCC had been ownership of a car that was 'five years old or more' which appealed to a number of members on the ground that, as their cars got older and threatened to disintegrate, they could replace them with a something suitably aged, and ensure that the club retained their loyalty.[37] As an argument, it had clear appeal but member Cecil Clutton, known as Sam, rejected it on the ground that the year 1930 marked the end of a distinctive phase of motorcar manufacture. 'From 1930 onward', Clutton argued, 'a new type of sports car had become popular, a type that was not particularly desired in the club.'[38] If his remark sounds snobbish, it was nevertheless only fair for a specialist organisation to remain true to its specialism, even if that meant redrafting some of its regulations. Tom, who supported Clutton, made the point that the VSCC filled the significant gap which existed between the VCC on the one hand and clubs that catered 'for modern cars' on the other. The proposed amendment was carried by twenty three votes to five.

September 1935 saw Tom elected to the VSCC Committee.[39] His attendance of meetings was somewhat irregular, but he earned the club's lasting gratitude by securing for its members a speed hill climb course.⁴⁰ As a 'riding mechanic,' he accompanied Carson in his Special and Clutton in the Itala on various occasions at Shelsley Walsh in Worcestershire. It was just the type of course that promised to suit VSCC members, but it was the 'exclusive preserve' of the Midlands Automobile Club.⁴¹ Aston Clinton in Bucks was dismayingly rough and although the club's Hartley Wintney neighbours Sir Denzil and Lady Cope made one of the drives of Bramshill House available for a trial, the offer was only for a single, specific event.⁴²

Then Tom remembered Prescott House, not far from Stanley Pontlarge, which had been the home of a Mr and Mrs Royds, friends of his parents. To reach the house, which stood near the top of a hill, a track wound its way up the scarp in a series of challenging bends, one so acute that when Tom visited in his GN he had to adopt the expedient of 'sliding the tail' – a manoeuvre in which the car's rear moved faster than the front – which had its perils, particularly since the shrubberies along the twisty route greatly reduced the available visibility.⁴³ By 1937, not only had Mr and Mrs Royds moved away but their successor Lt Col Godfrey Masters had also departed, leaving the house empty although the Gloucestershire Dairy Company purchased the surrounding estate.⁴⁴

When a number of VSCC members came to Worcestershire for the May 1937 Shelsley Walsh meeting, remembering the Prescott House drive with its tight bends, Tom persuaded Carson, Clutton and Bentley driver Forrest Lycett to venture south into Gloucestershire to try it out. Impressed by the experience and quick to see the route's speed-climb potential, they asked Tom to obtain an estimate for the cost of the necessary works on the road. Clutton, meanwhile, and a member named Kent Karslake explored the possibility of floating a company to purchasing the site on the VSCC's behalf. Notwithstanding a very 'reasonable' estimate for the work on the driveway, it was clear that the sheer ambition of the project promised to outstrip the VSCC's finances. 'The scheme', Tom recalled years after the discussions, financial and design-wise, had run their course, 'was too ambitious for a small and youthful club to undertake.' Yet the VSCC's initiative was not wasted. Clutton, who had friends among the membership of the wealthier, older-established Bugatti Owners' Club, brought all his diplomacy to bear upon persuading them to buy Prescott House to serve as their club house and using its drive, after modification, as their dedicated speed trial course.⁴⁵ It was agreed that the VSCC should hold their own event on it every year.

The venture also gave rise to one of Tom's first publications. A short story entitled 'New Corner', it appeared in *Mystery Stories* magazine in 1939 and shows supernatural elements at work within the matter-of-fact world of rally-driving.⁴⁶ The tale concerns the fictional Mercia Motor Club who have recently re-modelled their hill-climb course. Instead of gaining height through an easy 'series of zig-zag curves', in its modified form it ascends, straight and steep before taking a sharp left-hand bend – the 'new corner' of the story's title. But problems beset the route's alteration and the first meeting to take place at the re-modelled course proves cataclysmic with one renowned driver skidding as though on ice while another spins out of control at the same spot, his Grand Prix Rheinwagen ending up a 'mangled wreck' in the undergrowth while, distantly, club secretary Mr Nelson recalls 'some archaic superstition about a ring of old stones'.

The story ends with Nelson's ordering that the diabolical corner be enclosed behind a 'high and unclimbable fence', but this is no resolution to the stir that the corner's re-modelling has caused and questions linger. Why should the very terrain – familiar and, to all sight, stable beneath its tarmac surface – quicken in inexorable malice? What lies behind this clash between the spirits that inhere within the land and the people with their cars who disturb its ancient quiet? It is as though the conflict in Tom's thinking between the time-worn landscape on the one hand and the beguiling ingenuity of machinery on the other cannot always admit any simple resolution.

Chapter Six

Love and War

Tall, lean, good-humoured and possessed of a healthy sense of the ridiculous, Tom had no shortage of girlfriends. When he recalls his romantic liaisons in *Landscape with Machines*, it is usually with self-deprecating humour. In Stoke, he nursed a devotion to the golden-haired daughter of a family from Trentham – 'a goddess' – whom he took to the cinema on his motorbike and drove, on a wet day, in his leaky G.N., to a mutual friend's wedding in Hertfordshire. At Dursley, he had a liaison with the youngest of three sisters who lived up on Stinchcombe Hill. On his way to dine with them, Tom's G.N. failed, necessitating an *ad hoc* repair which meant that he arrived late, dripping with oil and generally looking as though he had 'come straight from the works'. Taking a 'beauteous young Hungerford maiden' for a weekend drive in his Humber, he encountered an officious off-duty policeman on the Wantage Road who tried to book him for the arcane offence of carrying a passenger in a vehicle running in trade plates on a Sunday. Since Tom had considerable familiarity with the more obscure provisions of the road traffic law and was a quicker thinker than the policeman, he was able to talk his way out of prosecution.[1] His tendency, in recalling each mishap, to focus more upon his means of transport than upon his companion gives the sense that none of these relationships was particularly serious. Then, late in 1935, he met a woman whom he credits with changing his life's course. In *Landscape with Machines* he calls her 'Anna', but at the time, he knew her as 'Cara'.

She appears to have adopted the Italian endearment as an alias – a form of disguise – and in her letters, she is generally careful to avoid giving any information which might identify her if they should happen to go astray. Once, admittedly, she writes to Tom from a verifiable London address – 56 Trinity Court, WC1, an art deco apartment block on the Gray's Inn Road – but more often, in place of an address, she supplies either the place name – 'Odiham' – or, as though teasingly taking her lead from some letting agent's advertisement, 'Vacant Cottage, Odiham'. Occasionally she uses hotel writing paper which, she may have reasoned, provided adequate cover. The frequent absence of dates in her missives means that any attempt to chart the development of the relationship must necessarily remain speculative. In one letter which she writes from 'The Orchard' – evidently the Windlesham home of friends of hers she refers to herself as a 'grass widow' and in another sent from the Windlesham address, she confides that her hosts are 'very sweet' to her. She adds that she is 'immersing' herself in 'the engrossing task of finding [….] a job' and that it will be a long time before she gets used to 'living alone and having no one to look after'.[2] In all likelihood, she was going through divorce proceedings.

How the two of them met, Tom does not relate, but she was, he asserts, older than him by some ten years.[3] Her precise identity has proved impossible to trace and Tom maintains a discreet silence on the subject. But he tells us that her interests turned chiefly upon literature and the arts; that she came from Irish stock; had a keen sense of humour; was bisexual; admired the essays of seventeenth-century lawyer and essayist Sir Thomas Overbury; worked intermittently on a biography of Queen Caroline; and that she 'could always be sure of finding acquaintances' at the Café Royal.[4] Ten of her letters to Tom survive.

Although he claims that he was never 'in love' with her, their correspondence casts some doubt upon his assertion that she was not in love with him.[5] A telegram that she sent on 28 November 1935, for instance, carries an unmistakable romantic charge.[6] 'DARLING' it reads, 'PLEASE PLEASE COME SLIPPING TOWARDS YOU CONTINUOUSLY PLEASE PHONE AS SOON AS YOU CAN'. Her letters customarily open 'My Love', 'My Darling' or 'Darling Tom'. At the end of December 1935, she writes that it is 'such bliss' to have him back, adding that she hopes that he 'will never know what it is to miss anyone' as much as she missed him over the previous week – in other words, at Christmas.[7] A letter of some two weeks later, concludes with this outpouring:

> Darling, you are in my heart and my mind, my rest and delight.
> You are with me in all I do and am, and I love you, Cara.[8]

Tom might not have loved her on the same terms, but that is not say that he did not value her companionship. Not only was she vivacious and amusing, but she had a genuine interest in his writing. Tom, of his own admission, kept the manuscript of 'Strange Vista' at this time 'securely hidden from every prying eye', as if it were 'outrageous pornography', but Cara somehow persuaded him to let her see it. That she had some knowledge of the workings of the literary world may have encouraged him to hand it over. When she had read it, she thanked him effusively. 'I know', she wrote:

> that one does not easily share anything as perfectly lovely and lovelily [sic] perfect as that […] In the last few hours I have read some of the most lovely passages in English prose that have come my way for a long time.

Not only did she promise to consult her publisher about it – 'It is of course, not his line of country at all', she warned, 'but he will know who would touch it, if any' – but she also volunteered 'to do all the typing of the script'.⁹

He accepted the offer, but it did not work out entirely well. Having typed some two thousand words, it appeared that Cara realised that the work was more demanding than she had expected. She left gaps in her typescript where she either could not read Tom's handwriting or felt 'uncertain' about his usage of what she took to be technical terms. She made no attempt to correct his mistakes in spelling and punctuation, on the ground that addressing such slips could wait until the novel was ready for proofreading and editing 'which', she perhaps misleadingly asserted 'need not bother us until after acceptance – if any!' She also ignored Tom's chapter headings, saying only that they would have 'to be gone over with a very great deal of careful consideration on our part'. Realising, perhaps, that she had done very little, she concluded her letter with a rather lame apology, writing:

> I do not get the time to work at this as I should like to and I do not wish you to think me a slacker, but with a few hours a day only to spare, it will take me some weeks to do it all.¹⁰

Whether she ever showed the manuscript to anyone in the publishing world is uncertain. Although Tom began to revise his novel, he found it impossible to write at the same time as managing the Phoenix Garage and eventually put his draft aside.¹¹

In the course of their relationship, he and Cara had a holiday in the Cotswolds together. They stayed in Chipping Campden, sunbathed in the fields, shared bread and cheese at a pub and visited Stratford-upon-Avon where the gusto of the Shakespeare Festival dispelled Tom's recollections of the fusty theatre trips that he had made as a schoolboy.¹² On 11 December 1936 – the evening when Edward VIII announced his decision to abdicate – the two of them saw Maurice Colbourne's *Charles the King* at the Lyric Theatre, Hammersmith and had the experience of rising at the end of the performance for the National Anthem only to see the orchestra hesitate and then pack their instruments away, 'realising' Tom writes, 'that at that moment there was no King for God to save'.¹³

Cara's letters suggest that she gave him ample cause for frustration. She could, on occasion, be manipulative and petulant. 'My dear child', she begins one of her undated communications, 'I can't bear to see you so depressed and miserable as you were last night…' Apparently he had given her a 'very severe telling off' because she had omitted to put on any make-up for his visits – something which she attempts in her letter to explain by saying that he would only see through the artifice. 'Today', she asserts with rather pathetic bravado, 'I fare forth to buy me a new paint-box and a Spanish comb set with rare jewels.'[14] It sounds like deliberate attention-seeking. Her two letters from Windlesham may have been the last that she wrote to him. One tells a farrago of a tale of what was probably a cat getting into her bed, and – a surreal touch – imploring her 'on Sunday morning [...] rather sloppily [...] to go to Madeira'.[15] The other concerns her search for a job and includes the sad sentence: 'I so hate this getting further and further away from you'.[16] Tom claims that it was she who eventually brought things between them to a close, writing 'an abrupt finis' to their relationship before either party began to lose interest in the other, but her reference to their drifting ever further apart sounds more like a protracted farewell.[17] Apparently they met again briefly after some ten years had passed. In strict date terms, it would have been sometime in 1946, which raises the possibility that he came across Cara in her native Ireland, when he was gathering material for *Green and Silver*, but the exact circumstances of this latter-life encounter are a mystery.

Towards the end of 1930s, Tom became increasingly restless. While he relished VSCC camaraderie and found the members' vehicles a constant source of fascination, he could not help but notice that the Phoenix Garage's finances were growing increasingly precarious. New development was changing the essential character of the ancient village of Hartley Wintney and meant that it felt neither like true town nor country proper, but an unhappy merging of the two. The way in which the song of nightingales in the Mildmay Oaks 'had to compete' with the blare of traffic noise from the A30 amounted, in Tom's view, to desecration.[18] A dangerous road, the A30 was the scene of frequent accidents. All too often, while lights shone and music played in the Phoenix Inn, Tom or Passini found himself in the garage's 1911 Rolls Royce Silver Ghost which the two of them had converted into a breakdown truck, heading to the scene of yet another crash, ready to set about the gruesome task of recovering the wreckage. Otherwise, they spent most of their time on minor repairs, small-scale maintenance jobs and the thankless business of pursuing defaulters. Often Tom's thoughts turned to the canals and – 'on hot summer days when traffic surged endlessly past the garage and the air was full of the stench of petrol fumes' – he would, yearningly, remember 'gliding along some narrow ribbon of still water between green fields' aboard *Cressy*.[19] He even hung up a waterways map on his bedroom wall and from his bed from which to plan potential journeys, thinking of *Cressy* with all the eagerness of the 'traveller in the desert [who] dreams of an oasis'.[20]

When Kyrle Willans moved away from Stoke after Kerr Stuart's collapse, he had sold *Cressy* to one E.J. Fortune – John Fortune, as he was known professionally – who was one of the editorial staff of the *Leicester Mercury*. Fortune was soon to marry and he intended that he and his wife Mollie, who loved boating, should live aboard *Cressy* at a mooring on the River Soar.²¹ Memories of the boat keen in his thinking, Tom had asked Mr Fortune to let him have first refusal if he were ever to decide to sell up.²²

At the time, it must have felt rather a forlorn hope but in 1936 Mr Fortune unexpectedly contacted him. His life had taken a tragic turn. Mollie had died earlier in the year and without her, he no longer wished to keep *Cressy* – the couple's one-time home. The news left Tom momentarily at a loss. On the one hand, to learn that *Cressy* was for sale counted for much. On the other, chronically short of funds, he could not afford to purchase the boat himself. Much to his surprise Kyrle Willans, whose career by this time had taken him and his family to a west country location completely out of reach of any navigable waterway agreed to buy *Cressy* back. With the boat in his ownership once more, Willans set out to Leicester with Bill to bring her south to Braunston Boatyard. Not long afterwards, he invited Tom and his motoring friend Aubrey Birks to join them for a 'reunion cruise' from Blisworth on the Grand Union Canal to Aylesbury and back. Tom also assisted in moving *Cressy* down the winding Oxford Canal when Willans decided that Banbury rather than Braunston should be her home port.²³

The minutes of the VSCC committee record that among the new members who joined in August 1937 were J.E. Bowen and 'Miss A Orrid' [sic].²⁴ John Edward Bowen, son of a baronet, and Angela Orred knew one another well for they had, in a sense, grown up together. Both their families had some footing in London and in 1925, when Angela had been ten and John seven, both families made their homes in the same part of Gloucestershire, Major Sir Edward Crowther Bowen buying Chesterton House in Cirencester and Major George Roland Orred purchasing Waterton House in nearby Ampney Crucis.²⁵ Over the years that followed, Gloucestershire newspapers often recorded the presence of Bowens and Orreds out fox-hunting, at agricultural shows and charity balls or attending the funerals of local dignitaries.²⁶ While John went to Eton, Angela acquired fluent French and Italian together with considerable skill in music and needlework. A debutante, she came out in the 1934 season; at the time, the *Tatler* described her as 'intriguing as well as lovely', the 'rosebud fairness' of her face complemented by 'a most determined chin'.²⁷

Tom relates that 'In the late summer of 1937' Angela swept into the Phoenix Yard in a white Alfa Romeo with all the panache of a character 'from a James Bond adventure'.²⁸ The Alfa, which she may optimistically have bought for participation in club events, came from the scrapyard, its surface glamour the result of her undertaking a 'very hurried and superficial face-lift' which had apparently 'included repainting it with what looked like whitewash'. But these details hardly detracted from the impact of her entrance to Tom's life. It may not precisely have been love at first sight, but the attraction, he says, 'was mutual'.[29]

The following November, Angela and John both competed in the VSCC's 1937 Cotswold Trial, albeit without much success; John's Chrysler attracted much attention but failed just beyond the 'observed section' while Angela's small MG proved unequal to the course's demands.[30] But whatever part John might previously have played in her life, from this time on, she drew increasingly close to Tom. Before long, he persuaded her to replace her Alfa with a 1920s racing Horstmann which they had discovered in a scrapyard near Swindon and which, to his mind, held far greater intrinsic interest.[31] She drove it in the 1938 Lewes Speed Trials where, apparently, its performance was 'mediocre in the extreme' although later in the year, it gave the couple an enchanted journey from Hartley Wintney to Prescott and they sped along the empty roads at 70 mph. In November, on a day of light frost, clear skies and bright sunshine, Angela joined Tom in his Humber for the Brighton Run. He was, by now, making 'increasingly frequent' visits to the South Kensington flat which she shared with her friend Margot.[32]

Restless, disillusioned with the garage business and uncomfortably aware that in all likelihood it would not be long before the country went to war, Tom's thoughts remained set upon the waterways. With hindsight, he may have come to think that in the context of engineering works, canals were unique in their capacity to harmonise with the landscape and even to enhance the delights of the natural world.[33] His more recent 'reunion' canal voyages aboard *Cressy* with the Willanses – going from Blisworth to Aylesbury on the Grand Union Canal and then taking her along the Oxford Canal from Braunston to Banbury – encouraged him to conjecture that he might reside very happily aboard a canal boat, travelling at will. Having drawn up hypothetical plans for the arrangement of the available space on board a narrow boat, he reasoned that it should be quite sufficient for 'civilised living'.[34] As for a career, he would seek to support himself by writing which struck him as 'the only occupation which could be reconciled with such a roving existence'.[35] He knew that it would not be easy, but at least he would not have either rent or rates to pay. Cara, on the strength of 'Strange Vista,' had encouraged him to think that he was not without talent. The popular magazine *Mystery Stories*, what was more, had recently accepted both 'New Corner' – his Prescott-inspired story – and 'The Mine' – a tale which sprang from his recollection of visiting the Snailbeach lead mines when he had been working in Stoke.

That he confided his plan or 'design for living' as he called it to Angela, and that she seized on it with zeal, says much about their growing fondness for one another something which John cannot have failed to notice.[36] Tom, out of caution, decided that Angela and he should test out their aptitude for living a shared life at close quarters and afloat and, much to her delight, in the spring of 1938, he took her for a week's trip on the Lower Avon Navigation. A neglected, semi-derelict waterway, Tom claims to have chosen it because, at a time when boats available for hire were scarce, he had secured the use of *Miranda*, a former lifeboat converted for use as a cabin cruiser, whose owner kept her 'at a mooring on the Warwickshire Avon at the village of Wyre Piddle'.[37] He may also have reasoned that by conducting their 'trial run' at living on the water in a remote part of the Midlands, they minimised any risk of John's encountering them unexpectedly. Certainly their voyage together went blissfully well, not least because Angela proved to be much more 'aquatically-minded' than Tom and showed no hesitation, when the need arose, in jumping off the boat and swimming ahead to close lock gates and to set paddles. By the week's end, she was every bit as enthusiastic as he about the prospect of living on the water, and the two of them were head over heels in love.[38]

Despite its lack of working capital, the Phoenix somehow continued in business over the summer and autumn of 1938. The garage's advertisement in the VSCC Bulletin of September 1938 lists for sale both a 1903 Humber, described as having completed four Brighton runs which sounds very much like Tom's car, and – suggestively – a Horstmann which may have been Angela's.[39] For both vehicles to have come onto the market simultaneously may – it is, admittedly, pure conjecture – reflect the dire state of Tom's finances of the time and Angela's tactful attempt to bail him out. Neither vehicle found a buyer, but by the end of the year the Phoenix Garage's need for a cash injection had become imperative.[40] Passini's family generously stepped in with a donation, but Tom, who had no money to spare, no wish to impose upon his parents and who was, anyway, making radical plans for his future in which the Phoenix would play no part, pulled out of the partnership. By now, Angela and he intended to marry. Since Lily and Mima knew of the couple's attachment, to them, the news would have been no surprise. Admittedly, the prospect of war, which was beginning to look inevitable, threatened to jeopardise their plans, but early in 1939 Tom persuaded Kyrle Willans, who was now living and working in Gloucestershire, to sell *Cressy* to him for £100. If the sum sounds small, it represented pretty much Tom's entire means of the time and years after the event he observed ruefully that 'few young men contemplating matrimony can have set up home on so little money'.[41] But his circumstances may have led him to drive a hard bargain. Bill Willans reckoned that negotiations over *Cressy*'s purchase led to relations between Tom and his father becoming somewhat strained.[42] The fact that Kyrle joined the Royal Naval Reserves for the duration of the war while Tom remained a non-combatant may further have exacerbated the tension between them.

Knowing that they would oppose it, for long time Angela did not tell her parents that she planned to marry Tom. When at last she broke the news, it caused immense trouble. The animus came largely from her father, Major Roland Orred, but since her mother deferred to him with unquestioning loyalty, there could be no question of her taking Angela's part.

Major Orred was a formidable adversary. If he had no aristocratic background, he did not lack wealth; his family were Cheshire landowners and when he had come of age in 1905, he had inherited extensive property around Runcorn, Weston and Tranmere.[43] Three years later, he married Cynthia Fitz-Clarence, great-granddaughter of William IV by his mistress Mrs Jordan, and relative of the Earls of Munster.[44] The First World War brought Orred swift promotion and he gained both a Military Cross and the *Croix de Guerre,* but he did not adjust easily to civilian life.

Driving between Alton and Petersfield in the early hours of 1 August 1930, for instance, he had managed to hit the car belonging to Flight Lieutenant George Symonds from RAF Tangmere. What was extraordinary about the accident was that Symonds had noticed the headlights of a vehicle approaching him at speed from behind and thoughtfully pulled onto the side of the road. But Major Orred's Rolls caught his rear wheel with such force that the impact not only thrust the airman hard against his car's steering column, but also propelled his vehicle into a ditch. A scraped side and lost front bumper notwithstanding, the Rolls hurtled on its way.

In time, there was a prosecution and Major Orred informed the presiding magistrates that Symonds had been 'a fool to stop on a corner'; Symonds stoutly denied having stopped in any such location. Under cross examination, the major told the court that he had been driving for twenty-four years; that he judged his speed at the time of the crash to have been about 40mph and admitted that he had given the hedge 'a pretty good bang'. The bench found him guilty of dangerous driving and fined him £4, at which point it emerged that he had two previous motoring convictions.[45]

The incident shows something of the measure of the man and in *Landscape with Machines* Tom gives a searing account of the bullying interview to which Major Orred subjected Angela and himself when they announced their intention to marry. He was also extremely rude to Mima when she called on him to put in a word on the young lovers' behalf.[46] 'He had the nerve to ask the Rolts for a settlement on me', remembered Angela years after the event when she met the journalist Ian Mackersey who was a great admirer of Tom's work.[47]

Yet there was another aspect to the story. For all the Major's swagger and boorishness, the life of the Orreds was not without its private grief. Angela was not their only child; there had been another daughter, Diana, who had died in December 1932 at the early age of twenty-three. For about a year after her death, the family effectively vanished from Gloucestershire's social life; exactly what effect the loss of their eldest daughter had upon them, we can only imagine. Outwardly, they appeared to adjust and Angela and her parents returned to the social scene in time for the 1934 London Season. Mrs Orred owned a house in smart Ovington Square, which gave them a footing in Belgravia and had the advantage of being close both to Angela's flat and the International Sportsmen's Club to which Major Orred belonged.[48] As the decade advanced not only did the Orreds and the Bowens decide to sell their Gloucestershire homes, but both couples moved to Sunninghill near Ascot – the Orreds to Sandy Ride House and the Bowens to Oakhurst.[49] It looks rather as though the Orreds' bereavement may have had the effect of drawing the two families closer together.

Late in 1937, Sir Edward Bowen died. It is possible that the Orred parents harboured some hope that Angela would marry John, who inherited both his father's title and his wealth. Assuredly such an outcome would have satisfied Major Orred's wish to see her 'marry into a title'.[50] More significantly, the Orred parents may have supposed that if Angela were to marry a man whose family they had known well over many years and whose mother was still their near neighbour, she would remain in closer contact with them than she would if she were to marry a stranger.

Marriage to a man who planned to lead a wandering life on the waterways was a different proposition entirely. For as long as she could, she avoided telling her parents about her relationship with Tom and when she could keep it secret no longer, the results were appalling. While Angela and her mother remained silent, Major Orred unleashed a vein of venomous snobbism. If we are to believe Tom, by dismissing his lack of money and prospects in such ripe phrases as 'shameful liaison', 'dirty garage mechanic' and 'infernal impudence' the Major not only ruled out all prospect of reasonable discussion but also thrust the couple onto their own resources.[51]

Tom's immediate response to the episode was to immerse himself in the practical aspects of making *Cressy* fit for residential use. While George Tooley of the Banbury boatyard and his sons attended to the boat's hull, he made the essential interior alterations.[52] For use as a holiday craft, *Cressy*'s fairly basic accommodation had sufficed. There had been no fewer than three twin-berth sleeping cabins, presumably diminutive in size, with a further two berths in the large saloon. There was also a small galley, a wash basin and a self-contained chemical closet. Tom determined to adapt and re-purpose these facilities to make the boat a hospitable home. He decided to keep the existing galley; provide *Cressy* with a dining saloon which could also do duty as a spare sleeping cabin; equip the large saloon with a coal-burning stove, comfortable chairs and built-in writing desk, and build bookshelves which would, in time, come to house Angela's clavichord. Besides a double bed, the state-room needed a bedside table, dressing table and hanging cupboard. Planning the bathroom called for special ingenuity since he was insistent that beside the washbasin there should also be a bath. Angela enterprisingly sourced and purchased a 'short bath' in London which promised, in theory, to be ideal for the limited space but which proved to be too wide to fit through the eighteen-and-a-half-inch doors. Undaunted, and having taken precise measurements, Tom bought a small, narrow bath from a Banbury ironmongers. Transporting it to the Boatyard required the services of three shop assistants with a handcart and caused much stir, but Tom helped the men to manoeuvre the tub on board and hoist it into position, high enough to allow the waste water to drain 'overside above water level'.[53] Its elevated position meant that there was ample space below it for a pair of good sized cupboards. The chemical closet, a relic of less luxurious days, remained on board, although rather than keep it amidships Tom moved it to a discreet location astern.

Precisely what triggered the incident in which Major Orred accused Angela of rank immorality, 'calling her a whore and a strumpet' and throwing her out of his house, Tom does not relate. It is no more than speculation, but if his account of the episode is to be believed, and he did not himself witness it at first hand, it raises the possibility that Angela thought that she might be pregnant and her parents guessed the source of her anxiety. Distraught at her father's outburst, she had driven overnight to Banbury boatyard and collapsed inside *Cressy* as soon as Tom opened the cabin doors.[54] Somehow, he managed to reassure her sufficiently for her to compose herself and head off on the following day to her London flat in resolute spirit. Mysterious as it is, the event may have persuaded the couple to bring their wedding forward. If so, together with George Tooley's falling ill at a critical moment in the course of working on *Cressy*'s hull, it would explain why, instead of sailing away on their honeymoon voyage as soon as they were married, the Rolts spent their first week of wedded life exploring the Upper Avon from a Bidford Inn.[55]

The wedding took place on 11 July 1939 in Caxton Hall, which served at the time as central London's register office. John Swainson, who, the previous year, had helped Tom to organise the VSCC's first Presteigne Rally was one witness; Angela's flatmate Margot was another.[56] They all dined afterwards at the Berkeley Hotel, but no-one was in celebratory mood. Besides the shadow which the family tension cast upon the atmosphere, war by this time was beginning to look unavoidable. It beggars belief, incidentally, that Angela would not at some point have told Tom about her sister.

But the rushed, fretful wedding had an unlikely coda. On 29 July 1939 the family notices of *Reading Mercury*, the local paper of the Sunninghill area, included an announcement of the marriage of Lionel Thomas Caswall Rolt, of Stanley Pontlarge, Winchcombe, Gloucestershire, to Angela Orred of Sunninghill.[57] The marriage took place, the notice states, 'at Westminster', a phrase which is as suggestive as it is vague. Assuredly its appearance did not mean either that Major and Mrs Orred had reconciled themselves to their daughter's marriage, or indeed that they were ready to welcome Tom into their family. Although his claim that they cut themselves 'off from her for good' is not strictly correct, the only communication that Angela had with her father over the succeeding years came when he tried to siphon off the monies that made up her inheritance, although she did occasionally meet her mother. But the damage was done and Tom was convinced that her parents' behaviour around the time of her wedding left her bearing 'an intolerable psychological burden' the consequences of which he never fully grasped.[58]

Chapter Seven

Pacifism and the Path to Fame

From Banbury, Tom planned that Angela and he should head towards the Welsh section of the Shropshire Union Canal to fulfil his ambition of taking *Cressy* across the Pontcysyllte aqueduct. At last the Tooleys finished their works on the boat and the couple left Banbury aboard *Cressy*, her new paintwork gleaming, early in the afternoon of 27 July 1939, travelling north through the meadows along the Oxford Canal. For this first leg of their journey, old Mr Tooley from the boat yard joined them, ostensibly to help with the locks, although in truth it was more for pleasure than from any necessity. He disembarked at Cropredy where they moored for the night. On their first day's travel, they had covered four and half miles.

Angela and Tom went north by leisurely stages and on a typical day, they would cover something between three and eleven slow miles. On their way, they paused to take the tram into the centre of Leicester, they detoured through Loughborough's backstreets to visit John Taylor's' bell foundry and at Burton-on-Trent they stopped for a tour of Allsop's brewery. They bought milk and other provisions from farms along their route; paused often to explore ancient churches and historic houses and joined in the conversation and singing that they found in the canal side pubs. If it sounds an idyllic journey, the growing threat of war and recent clash with Angela's father haunted the couple's thinking. These 'two subjects', Tom remembered years later when he came to write *Landscape with Canals*, they 'never discussed, but could not forget'.

He kept a record of their travel in a board-bound notebook – the *Cressy* log.[1] Here he wrote a succinct account of each day's progress, complete with running tally of the daily and total mileage travelled, and the number of locks and tunnels negotiated. Factual and unvarnished, its entries record the times of departure from each mooring and of tying up at the close of the day's travel. Often, especially in the earlier entries, he also mentions the weather; supplies purchased; vessels and wildlife encountered; mechanical and navigational mishaps, and the places which Angela and he visited. Although ships' logs generally confine themselves to factual matters of navigation, Tom often digresses to record his more memorable encounters and exchanges. Near Bedworth on the Coventry Canal, for instance, he met an elderly boatman whose house had been scheduled for demolition under the local authority's slum clearance plan. Asked to vacate the building, the old man had declined, telling the housing officer that he and his sister 'would walk straight down into cut, sooner than leave the house where we was born'.[2] It was just the sort of anecdote that Tom enjoyed; he always appreciated tales of cussedness in the face of petty officialdom. On the page, he can be extremely outspoken. He and describes the stretch of the Trent and Mersey Canal between Dallow Lane and Branstone as 'the dullest reach we have been through so far'; dismisses Rugeley as 'a squalid, dirty little town, set in country worthy of better things' and considers Wolfhamcote Church – he spells it 'Wolfamcote' – to be in a 'very bad state of repair'.[3] When the *Cressy* log mentions Angela, she is often buying provisions or cooking in the galley. Tom's tendency to refer to her as 'Mate' – he is 'Captain' – carries the chauvinism of its time; early in the voyage he records, for instance, that 'Mate took over the helm for the first time and handled the boat very ably'.[4] When Ian Mackersey visited Angela in France, she added some substance to the *Cressy* log account. Tom, she remembered, taught her the elements of navigation so that the two of them 'worked the boat together as a team'; given the tiller, she steered carefully and well, never running into anything and never putting the vessel aground.[5]

The chief annoyance that the two of them encountered in the early phase of their journey, came from gongoozlers – in canal terminology, people who stand 'staring for prolonged periods at anything out of the common'.[6] On 10 August, the Rolts moored in Blaby on the River Soar to stock up on provisions and have a drink in the County Arms. In their absence, some particularly brazen gongoozlers actually boarded *Cressy* and had 'to be more or less forcibly ejected' when the couple returned.[7] The next day, they joined the Grand Union Canal's Leicester Line and journalist John Fortune – *Cressy*'s former owner – called on them at Aylestone, keen to see his old boat again. He appears to have arranged for the local paper to include an item about their travels at the same time.

'Couple's novel journey' announced the *Leicester Daily Mercury* of 11 August with a photograph of Tom and Angela on *Cressy*'s deck; 'To North Wales by Canal will be the unique form of transport favoured by a young married couple, Mr and Mrs L.T.C Rolt…'[8] Explaining that they intended to reach Llangollen by September, the piece launched into an enthusiastic description of *Cressy*'s colour scheme and the on-board 'bathroom and electric light' before commenting that the Rolts apparently found living on the canals 'much more attractive than life in a humdrum corner of suburbia'.[9] Predictably, the item caught much attention and a crowd of people arrived to inspect the boat for themselves, 'asking "how far to Wales?" and such like nonsensical questions' noted Tom caustically in the *Cressy* log.[10] In view of the newspaper's remark about 'humdrum corners' it was ironic that the next section of the Rolts' route took them 'under innumerable bridges', past a hinterland of mills and factories, rubbish dumps, and the town's gas works and power station. A detail that he does not mention in the *Cressy* log, but which he would recall when he came to write *Narrow Boat* was that Belgrave Lock, situated just beyond Wolsey's mills, was 'full of dead rats'.[11] It was not a vision of industry from which even Tom could derive any pleasure.

On 14 August, they rejoined the river, passed the village of Mount Sorrel, marking the immense granite quarries on the left bank, and moored up in Barrow on Soar. It was an attractive spot; the Navigation Inn offered 'excellent' beer; the dabchicks in their black summer plumage 'with bright chestnut cheeks and throat' were a delight and Tom enjoyed seeing Beyer Garratt locomotives hauling coal trains on the old Midland Railway mainline.[12] While they were there, Tom bought a newspaper in which he read that John Bowen had crashed his Maserati at Donington Park and died instantly.[13] He had held his baronetcy for less than two years.

Tom makes no mention of John's death in the *Cressy* log, but writing *Landscape with Canals* some thirty or so years after the event, he admits that John had been 'a recent ex-boy-friend of Angela's'; that his marriage to Angela had hit John 'very hard' and that both Angela and he 'felt in some degree responsible' for John's death.[14] They are grave disclosures and point to a tale which, in *Landscape with Machines*, Tom tells only in part. As to what conversation ever passed between Angela and himself concerning John Bowen, he gives no indication whatever.

Country-wide, the Donington calamity caused much stir.[15] At the subsequent inquest, a witness named Michael Paling Smith gave evidence to the effect that Bowen had been about to overtake another car when he hit the grass verge which caused his Maserati to swing round, crash sideways through the barrier, overturn 'three or four times' and eject its driver. The coroner, a Mr H.J. Deane, remarked that Britons often relished sports which contained an 'element of danger' and added that he was 'sure that the jury would not like to say anything that would check anyone's sporting liberties'. Following his gist, they returned a verdict of accidental death.[16] A week later, the Rolts' route along the Trent and Mersey Canal bought them within sight of Donington Park. Tom says nothing about their emotions at the time, but remarks only that they both accidentally left their windlasses behind at Swarkestone Lock.[17]

On 1 September, while they took on petrol at Trentham Bridge, a passing horse-boatman told them that Hitler had invaded Poland.[18] Although Tom and Angela must both have recognised the significance of the news, they refused to abandon their journey and continued on their course through Stoke. The canal took them past Kerr Stuart's old site, Tom sadly marking the state of dereliction of the California Works with the glass in all the windows smashed. Once the tug had hauled them through the Harecastle Tunnel, they had moored for the night near Church Lawton. Although there had as yet been no formal declaration of war, in compliance with Air Raid Precaution legislation there was a blackout in operation and no after-dark street lighting. Did the Rolts but know it, Stoke had already appointed A.R.P. wardens and appealed for volunteers to fill sandbags and distribute gas masks, while the town's drapers had seen a great 'run' on dark material for use as blackout curtains.[19] Two days later, from a mooring near Middlewich, the Rolts heard Prime Minister Neville Chamberlain broadcast to the nation, announcing that Britain was at war with Germany. No longer could the couple cling to the hope that peace might prevail. Angela, stricken by John's untimely passing and reeling after the breach with her parents, wept and Tom could find no words to comfort her.

But as the hours passed, they made plans. There could be no question now of their reaching Llangollen and the book that Tom was hoping to write would have to wait. Although he had no intention of enlisting – indeed, Ian Mackersey relates that Lily wrote to Tom specifically begging him 'not to volunteer for military service' – he recognised the importance of contributing to the war effort.[20] Since engineering was a reserved occupation – that is to say, one which was considered sufficiently important to exempt him from enlisting in the armed forces – he decided to seek an engineer's post for the duration of the hostilities and serve the country in a civilian capacity.[21]

The concept of pacifism is not something that Tom cares to discuss, but his antipathy to war appears to have owed as much to an instinctive repugnance as it did to reasoned argument. At one time, he had tried to set out his views in an essay entitled 'This modern Generation – A Response to its Critics'. Unpublished, it survives in manuscript form, written in tidy script, as though he had some idea of preparing it for submission to a likely periodical. Although it carries no date, its glancing references to Fascism and Nazism suggest that he may have written it in the early to mid-1930s. In it, he argues that his generation have no time either for established Anglican Christianity or, significantly, for Empire. The one he equates with 'hypocrisy and mysticism'; the other, he dismisses as the outcome of unbridled bellicose chauvinism. 'We prefer our country to others because we happened [sic] to be better accustomed to it', he reasons. 'At the same time, we have no cause to believe that we and our particular little portion of the globe are any better than any others....' For the promises of political leaders, he has little time. 'We know only', he writes, 'that the present system is outworn and tottering. That is why we fly to the support of any new movement that promises action rather than stagnation. Hence the support of Fascism, and in Germany, the sweeping success of the Nazi regime.' At the same time, he deplores the 'narrow, nationalistic bias' which tends to characterise such movements. 'At heart', he insists, his generation are not nationalists, but 'the first terrestrials, though we do not realise it.' By his use of the term 'terrestrial' – he takes it above and beyond its usual sense of something that is 'of earth' – Tom appears to try to evoke the idea of humanity's primary allegiance being to the rest of humankind, rather than to fellow nationals. Unfortunately, 'terrestrialism' and its implications are not concepts that he cares to explore. If, as a personal manifesto 'This modern Generation' feels rather underdeveloped, its ambition is unmistakeable. Caught on the cusp of war, Tom sought to remain true to the ideals that he had tried to express in his monograph. The fact, incidentally, that he retained the essay suggests that, although he does not appear ever to have finished it, he valued it as the record of his thinking at a significant stage in his life. Whether he ever had any thought of revising the piece to refine its arguments it is impossible to say.[22]

Musing over the question of where he was to find likely civilian employment in September 1939, an answer quickly presented itself. The Merlin engines which powered Spitfires and Hurricanes were in production at Rolls Royce's Pym's Lane factory on the outskirts of nearby Crewe. There was, Tom reasoned, every chance that Rolls Royce would have an opening for him. Two days later, having found a provisional mooring at Nantwich, he presented himself for interview and produced the certificates confirming completion of his apprenticeship and testifying to his capabilities. Immediately he found himself offered a fitter's post with instructions to be ready to start work the following Monday and to join the Amalgamated Engineering Union. To discover that membership of the Amalgamated Engineering Union was compulsory annoyed him somewhat, but he bowed to the inevitable. Besides, more pressing matters claimed his attention. Angela and he had been quite unable to manoeuvre *Cressy* into Nantwich's silty basin and they reluctantly decided to move her to Church Minshull. Unlike Nantwich, it was not on a bus route, but Tom bought himself a second hand motorcycle together with a bicycle for Angela.

Despite his irritation at having to join a trades union, Tom looked forward to being back in regular work and, what was more, for rather an august firm. There would, he reasoned, have been changes in workplace practice since his brief spell of employment with Thornycrofts of Basingstoke in 1933 – it had been the last time that he had worked for an engineering firm of any size – but he was confident that working for Rolls Royce would bring fulfilment and job satisfaction.

Under the glare of the assembly shop's mercury vapour lamps, his respect for the company evaporated fast. Rolls Royce had acquired the site a year previously in response to the Government's wish to increase capacity for aero-engine manufacture. Before that, it had been farmland.[23] Tom found the state-of-the-art factory loathsome; it had no windows; the lights made his workmates look livid and sickly while the stench of stale sweat and cigarette-smoke hit him every morning 'like a blow in the face'.[24] Rashly, he had assumed that the experience to which his certificates attested would lead to his being allocated the satisfying task of fitting up Merlin engines from scratch, much as he had built up larger and more complex types of diesel engine for Listers. On his first day, it emerged that his sole task was to tap holes for studs in 'an endless succession of Merlin cylinder blocks' as they appeared before him on a conveyor. In his view, it was a task which might have suited a trained ape.[25]

Disagreeable and boring it may have been, but his protests feel somewhat disingenuous. The absence of windows precluded any inadvertent contravention of blackout regulations. As a newcomer, it was only to be expected that he should start his employment at a relatively lowly level. Admittedly his job was repetitive – it sounds tedious beyond words – but he missed an important point about Rolls Royce's wartime production methods. It was in the company's interest to enable novice recruits who had no engineering skill or background to learn how to perform a given task in a short time. By standardising work practices and using machinery to perform the work which would once have been the preserve of seasoned craftsmen, the company could entrust intricate operations to an inexperienced workforce with minimal risk of mistakes occurring. Although Tom complains about workplace 'boredom and apathy', arguably Rolls Royce had equipped itself rather deftly to the meet the demands of the time. When, rather than enlist, Tom took up his post with the company, he was entirely within his rights, but, at the same time, blatantly out of tune with the mood of the moment. Crewe had been quick to put itself on a war-footing. Even before the introduction of rationing, the *Crewe Chronicle* bracingly urged its readers to grow their own vegetables, stating that 'The best way to convert a lawn into a kitchen garden is by trenching' and sternly warned them not to waste paper but to 'conserve supplies'.[26] No doubt much conversation at Rolls Royce turned upon the war effort and it is likely that Tom came in for some pointed questioning from his workmates as to why he had not enlisted. As much as the lighting, atmosphere and monotony, it may have accounted for his dislike of the job.

Then, in October 1939, his old boss at the Aldbourne Engineering Company wrote to ask whether he might be available to work for them again. Aware that in order to feed the nation, they had to produce higher crop-yields with less help on hand, the farmers found themselves increasingly reliant upon machinery. Like many other small agricultural-engineering concerns, Aldbourne soon had more work than it could manage. Although Tom's previous spell of employment at Aldbourne had been brief and ended abruptly when the job 'died' on him, while it lasted, it had been rewarding. He had serviced engines of varying degrees of antiquity all across Wiltshire, gained practical knowledge of most farming operations and travelled the length and breadth of the great chalk downs. He had come to know Crofton Pumping Station, whose beam engines supplied water to the summit level of the Kennet and Avon Canal, and befriended the elderly engineman who maintained them. The wage that Aldbourne offered was less than half of what Rolls Royce paid him but he handed in his notice without a qualm.[27]

His mind made up, on 3 November 1939 he broke his two month silence in the *Cressy* log with the terse entry:

> Having left Rolls and resumed writing, also because of a leak
> [...] and other reasons, we decided to move south.[28]

Recognising the inevitability of wartime petrol rationing, while *Cressy* remained at her Church Minshull mooring, he had resourcefully adapted her Model T Ford engine to run on paraffin by fitting a vaporizer. It meant running a fuel line from the tank which held paraffin for the lamps and stove to the carburettor, but that presented him with minimal difficulty.[29] Needing now to purchase petrol only in the small quantities required for starting and warming up the engine, Angela and he were ready to embark on the next phase of their travel.

The Rolts' journey from Church Minshull to Banbury where they moored up in Tooley's Yard took them from 11 November until 7 December 1939; that is, rather less than a month.[30] Travelling north in high summer, their journey had occupied almost six weeks but then, the couple had been at leisure. Mindful of the work that awaited Tom in Aldbourne, on their return south, they travelled faster, making fewer diversions. Besides, the bleak autumn weather portended a harsh winter. Their route took them by the Shropshire Union Canal to Autherley Junction where they followed the Staffordshire and Worcestershire Canal to the Trent and Mersey Canal at Great Haywood. At Fradley Junction they joined the Coventry Canal, pausing at Huddlesford to detour (for a second time) into Lichfield. Resuming their travel, they headed to Hawkesbury Junction and headed by the Oxford Canal to Banbury. Autumn was by now giving way inexorably to winter; having moored for the night near Cropredy Lock, when they returned from an evening at the Red Lion, they found the canal frozen over.[31] The following day, Tom had to thaw *Cressy*'s water pump before he could start the engine. The freezing conditions confronted him with a dilemma. To continue might prove perilous; as he observed, 'thin ice cuts the timbers of a boat badly'. At the same time, he did not care for the idea of spending the winter iced up in Cropredy and had no idea how long the freeze would last. Making the short journey on to Banbury was, he decided, a risk worth taking. In the event, although the freeze made opening lock gates on their way difficult, beyond Banbury Lock, the discharge of warm water from the condensers from the nearby aluminium works ensured that the canal was clear of ice. Soon,

Angela jumped ashore to raise the drawbridge gates that led into Tooley's Yard and at 2 p.m. they moored up. They would remain there – did they but know it – until the following March.

In this quiet and otherwise fallow time, Tom resumed work on the draft of his book. He had put it aside during his time at Rolls Royce, perhaps unwilling to revisit the pleasures of canal travel when he inhabited such a different world for the greater part of each day, but now he seized the time to commit himself in earnest to his manuscript. It concerned his travel aboard *Cressy* and he gave it the title 'A Painted Ship'. Although it dealt with canal life rather than the life of the land, Tom may have had memories of *Shepherd's Country* – H.J. Massingham's book about the Cotswolds, a copy of which Mima gave him shortly after its publication in 1938 – alive in his mind as he wrote.[32]

Massingham lamented what he saw as the decline of Cotswold rural life and his brief account of Gloucestershire signposting gives some flavour of his sense of loss. 'The old signposts', he explains:

> gave you the direction from the village to village, often with an obliging finger and the number of furlongs thither thrown in. The modern ones point the way to large towns from twenty to fifty miles away. There is a world of tragic meaning in that change.[33]

Doing away with old finger posts might seem a trifling matter, but to Massingham, the changed assumption about the likely places to which people might wish to travel – privileging the national over the local – was a shocking new species of geography in which small settlements counted for nothing. Massingham championed traditional craft and deplored its decline. He writes disparagingly, for instance, of roofs constructed from 'old stone' but cut, by machine, to precise lengths, the surface texture smoothed away, robbing the finished roof of all character.[34] He liked the way that the terms of 'chad' and 'chase' survived for a pickaxe and wedge, since 'to give a tool a Christian as well as a generic name helps to keep the [quarryman's] job still human and personal and to save the trade from degenerating into mere commerce'.[35] Tom valued such sentiments highly.

To write about canals at the same time that Angela and he were on a canal voyage had clear logic and although Tom's day-to-day logbook entries are often brief to the point of being skeletal, they nevertheless served him as an *aide-memoire*. Decades after *Narrow Boat*'s publication, in her conversation with Mackersey, Angela recalled Tom's writing routine. At the end of each day aboard *Cressy*, she remembered, he would go to his writing desk to write up the log before turning to the draft of the book-in-progress that eventually became *Narrow Boat*. As a wedding present, she had given him a typewriter and she remembered him typing swiftly 'with two fingers'. While he worked, she read, sewed or played the clavichord which Tom's friend Cecil Clutton, known as 'Sam', from the VSCC had found for her. That the book follows a clear taut arrangement, its chapters compact and self-contained, each centring on a specific place or event, suggests that he composed it along methodical, orderly lines. Although his text expands the log book record, he nevertheless remains faithful to the framework that it provides.

Reflecting Tom's mood of the time, it is often an angry book. Impoverished and anxious, the nation went into war after years of economic slump. It had left a legacy of neglect and decline. His entry in the *Cressy* log for 2 August 1939, reads 'After tea, walked into North Kilworth, a large rambling village of no particular merit. Tragic to see the number of thatched cottages re-roofed with corrugated iron....'[36] In *Narrow Boat*, the melancholy gives way to open wrath. Deploring what he views as the tendency of the south Leicestershire countrymen to eschew their rural roots 'for the get-rich-quick lure of the industrial towns', Tom describes what he terms 'this new generation of "countrymen"' with unconcealed scorn, acid in his observation of their return home after a day's work in 'some factory or other, [....] their tin lunch-cases tucked under their arms'. They had, in his view, 'broken faith' with the land and the scars of its neglect – 'untrimmed hedgerows, choked ditches and gates drunkenly leaning' – are all too plain to see. 'The rural arts', he writes bitterly:

> had gone with the old rural community and the village was dispirited and dead, the only new life a row of unsightly new bungalows which had sprung up along the main road conveniently near the station.[37]

At intervals, Angela remembers, he hit an impasse. Uncertain what to say or how to say it, he would sit motionless, 'hunched over the typewriter chain-smoking' until, without warning, the language fell into place once more and he typed on in full fluency. When he had finished for the evening, he would read her what he had written. She remembered being 'struck by the beauty of the prose' but at the same time – secretly – could not imagine how a book about life on the canals would captivate and delight the general, book-buying public in the way that it so entranced her husband and herself.[38]

Tom's bosses at Aldbourne evidently took a relaxed view over his start date. Having over-wintered at Banbury, it was not until 1 March 1940 that the Rolts resumed their journey south. Travelling by the Oxford Canal, the

Thames, the Kennet Navigation which showed all the marks of neglect and the Kennet and Avon Canal which was nearly derelict, they had a difficult passage, their progress hampered by underwater obstructions, hidden shoals and leaky lock-gates.

On 12 March, they came through Kintbury where they stopped at the canalside inn for lunch and took on coal near the lock. Although the lock-keeper was old and deaf, he helped them through Fowles and Brunsden Locks.[39] At Dun Mill Lock they took on board his Hungerford counterpart who helped them through Hungerford Swing Bridge. They moored at nearby Wooldridge's Wharf amid steady rain, and there *Cressy* would remain for the following year.

Except to describe his work as 'varied and interesting', Tom says relatively little about his second period of employment with the Aldbourne Engineering Company. Finding himself despatched to the 'binder-school' at Massey-Harris's Trafford plant outside Manchester, he soon realised that, far from any offering practical instruction in the workings of a new and intricate piece of machinery, the manufacturers had cunningly provided themselves with a source of cheap labour. Unimpressed, Tom – together with men from various other agricultural engineering contractors – found himself assembling new binders from scratch, the discrete components having arrived at Trafford from the US by way of the Manchester Ship Canal. It was neither an edifying nor an enjoyable experience and he returned south knowing no more about power-driven reaper-binders and the operation of their delicate knotter mechanism than he had previously. Even so, he soon found himself regarded as the Aldbourne 'binder expert' although he reckoned that his capacity to resolve problems with the machinery owed more to 'low cunning and perseverance' than it did to engineering skill.[40]

Work on his book had occupied Tom throughout the winter of 1939–40, when ice confined *Cressy* to Tooley's Yard and by the time that Angela and he set off for Hungerford, he had 'completed all but the last chapter or so'.[41] Despite Angela's recollection of his typing in the evenings, an early version of it survives drafted in longhand on thick foolscap paper, ruled and with a broad left hand margin.[42] At some point the handwritten version of the text came in for much revision – it is heavy with arrows signalling changes to the order of paragraphs, for instance – and in time, putting his wedding present typewriter to solid use, he prepared a fair copy typescript.

By February 1941 he was in correspondence about the work with publishers George Allen & Unwin Limited, more often known simply as 'Allen & Unwin'. Philip Unwin, nephew of the company's founder and active in its editorial department, wrote to Tom at Wooldridge's Wharf asking to see the *Painted Ship* manuscript. He had, he explained, read and enjoyed A.P. Herbert's 1930 novel, *The Water Gipsies*, and Tom's canal book had caught his interest. Not only did he ask Tom to send him the manuscript but at the same time sought his views on the book's likely sales potential. Tom's response to this enquiry does not survive, but he evidently despatched the manuscript, which Philip Unwin clearly liked. But war time austerity had hit the publishing industry hard. If Allen & Unwin were to bring out 'A Painted Ship', he explained, Tom would have to finance the publication. Tom, understandably, was less than happy with this proposal, and Unwin, to his credit, returned the manuscript, acknowledging Tom's reservations about putting up the money in an understanding note. Around the same time, Tom received a letter from Harry Batsford, director of fine arts and technical publishing company B.T. Batsford Ltd. In it, Batsford mentions 'concessions' which Tom must have offered in order to see the book published; it is possible that this detail lies behind his memory of a publisher who offered '£75 outright' for the work. But Batsford was slippery. Quick to tell Tom how highly he thought of the book, at the same time he demurred over the question of quite how popular it was likely to prove with the reading public. While in his view, the war had not led to any major falling-off of book sales, 'the more special a book is' he asserted, 'the less it shares in the mild current popularity'. Indecision hangs heavily upon his letter. Having asked Tom to let him see the manuscript again, in the same paragraph he remarked that the two of them would need to agree upon 'a title of really strong popular appeal'. From his letter, it is less than clear whether they had a firm deal or not. If the twists in Mr Batsford's thinking made Tom suspicious, it was no wonder that he declined to publish 'A Painted Ship' with Batsford. He also approached J.M. Dent & Sons, publishers of the Everyman Library. and Ernest Franklin Bozman, who was both on Dent's editorial staff and an editor of *Everyman's Encyclopaedia*, not only liked what he read but genuinely believed that it had enough viable commercial potential 'to do well enough to pay for itself'. Even so he declined it, citing 'practical difficulties' by which he may have meant wartime paper rationing which, he explained, was less than favourable to the publishing industry.[43]

Although disappointed, Tom cut his losses. Not only did he begin to work on a second book, *High Horse Riderless*, but he also started writing articles for literary-political magazines and specialist journals. Their subjects ranged from set design in the theatre through motor car production to humanity's changing relationship with the machine.[44] Early in 1941, having been at Aldbourne no longer than a year, he realised that for all the satisfaction of his work, it did not produce a living wage. The 'slender allowance' which Angela rather surprisingly received from her parents might, he reasoned, cease at any moment upon Major Orred's whim. At the instigation of Harry Rose, who was one of the few friends he had made at Cheltenham College, he successfully applied for a position in the Ministry of Supply which promised greater economic security than his Aldbourne post. He would become a Technical Assistant, undertaking what seems to have been a quality-control function in which he checked that vehicle spares supplied both to the armed services and for essential civilian use were of the required standard.⁴⁵ He would, he realised, have to travel to such far-flung places as Grantham, Chesterfield and Shrewsbury. Having moved *Cressy* to Banbury for repairs and refurbishment in the boatyard, Tom, who had 'laid up' his Alvis for the war, borrowed his father's four-seater model for work. Often, it proved a great asset. 'It was not', he remembered, 'the type of vehicle that engineers normally associate with "the man from the Ministry" and on this account they were the more ready to be friendly.'⁴⁶ Notwithstanding the impression that it made, he had no wish to wear it out in Government service and, in time, he replaced it with an Austin 'Big Seven' saloon which, although – or, perhaps, because – he thought it 'a terrible little car in every way', seemed more fitting for his occupation.⁴⁷

In June 1941, Lily had a stroke and although Tom and Angela hastened to Stanley Pontlarge, he died before they arrived. Tom found his death distressing in ways which he had not foreseen. Arguments in which he had taken his mother's side against his father; his father's 'interminable and oft-repeated stories' which he had frequently and impatiently interrupted and enthusiasms like angling in which he had doggedly declined to partake and all the affection which, somehow, he had left unexpressed – came back to him. Hearing the inexorable ticking of his father's watch, he did not feel grief, so much as 'remorse and self-reproach'.⁴⁸

By the end of August, the work on *Cressy* was complete. Since the Ministry of Supply, by this time, had indicated that Tom's work would largely confine itself to 'the area around Redditch, Birmingham and Wolverhampton', having scoured the maps for an attractive and well-located country mooring, the Rolts settled on Tardebigge on the Worcester and Birmingham Canal. Tom arranged for Mima to join them for the voyage north, and took a week's leave. They reached Tardebigge New Wharf at 12 noon on 4 September 1941.⁴⁹ It would be their home for the following five or so years.

On leave in the summer of 1943, encouraged by the fact that he was now earning a substantial wage, Tom took Angela to Wales for a holiday. They stayed at Minffordd near Cader Idris, and one bleak day they took the bus into the modest seaside resort of Towyn – now known as Tywyn – bound on a mission of totally unforeseeable significance for Tom's future. Their intention was to make their way back to their holiday lodging by taking the small, independent Talyllyn Railway to the village of Abergynolwyn and then covering the final five or so miles on foot. But fate scuppered their plans. When the Rolts arrived at the Wharf Station on a damp day, they found a notice bearing the stark statement 'No Train Today'. The only locomotive on the Talyllyn line that was in anything resembling working order at this time had broken down and was undergoing essential repair. Since the only return bus of the day had already left, the couple resigned themselves to walking along the line for seven miles to the terminus and then following the road to Minffordd. As it turned out, just beyond Abergynolwyn they met a friendly district nurse on her rounds and she gave them a lift to their destination.[50]

About his work at this time, Tom says understandably little, beyond mentioning the need to fill in a weekly form to claim his travel and subsistence allowances.[51] His 1946 short story 'Hawley Bank Foundry' with its sharp portrayal of the overbearing business man, George Frimley, may give some idea of his opinion of the managers of the various works within the area to which the Ministry had allocated him. He found them more congenial than Rolls Royce, but once his wartime employment ended, he left the Ministry with what Angela terms 'whoops of delight' and exchanged his formal garb for 'the old corduroys'. Husband and wife marked the occasion by burning 'PILES of papers – Official, secret and all sorts' and Tom was quick to sell the despised Austin and retrieve his 1924 'duck's back' Alvis from the Stanley Pontlarge shed.[52] Nevertheless, his war work earned him warm commendation. Among his paper is a letter on Ministry of Supply headed paper, thanking him for his help and assuring him that his 'excellent efforts' were appreciated.[53] Sadly, the letter writer's signature is indecipherable.

By this time, Tom's journalism brought him to the attention of other writers, among them his hero H.J. Massingham, with whom he began to exchange letters. In the course of their correspondence, Massingham suggested that Tom might care to write 'something about the canals' and when Tom told him about the fate of 'A Painted Ship,' the manuscript now consigned to a suitcase under the Rolts' bed, Massingham demanded to see it. Not only did he read and enjoy it, but – conscious of the influence that he exercised in 1940s literary circles – he decided to take matters into his own hands and approach Douglas Jerrold of Eyre & Spottiswoode on Tom's account. Not long afterwards, Tom received an 'encouraging letter' from Jerrold to be followed, in September 1943, by a contract.[54]

Disliking the 'arty-crafty' associations of his original title Tom decided to rename the book *Narrow Boat* which, to his thinking, had the merit of directness. He deplored the publishers' reluctance to use Angela's canal photographs to illustrate the book and never reconciled himself to their decision to commission scraper-board images by Denys Watkins-Pitchford instead, complaining that they would chime more with a 'hey-nonny-no, under-the-greenwood-tree type of text'.[55] Despite his reservations, when the book appeared in 1944, produced unassumingly in compliance with the War Economy Standard, it was an instant success. Tom had known the canals for fifteen years and lived on them for five. Soon, he would find himself fighting for their survival.

Chapter Eight

Robert Aickman

If *Narrow Boat*'s popularity surprised him, Tom was quick to make use of its advantages. Even before it came out, he secured through Massingham's influence a contract with publisher Robert Hale to write a book for their English Counties series.[1] He chose to write about Worcestershire in part because, based at Tardebigge, he was well-placed to explore the county while travelling for his work. But Worcestershire's history and terrain, which embraced both agriculture and industry, put the questions which so much interested him – questions about 'the interdependence between man and nature' and specifically about humanity's influence upon the landscape and the land's influence on its inhabitants – into clear perspective. Thrilled to have so engaging a commission, he happily absorbed himself in gathering material for the project.

It was not long before Tom accepted an invitation from one Henry Cornelius, a film producer, to act as technical adviser to *Painted Boats* – a film which Ealing Studios were making. He soon came to find the business of film making tedious beyond words and dismissed its superficial glamour contemptuously as 'tinsel', but he relished such tasks as selecting locations, choosing suitable boats and providing background information about canal history; apparently he also coined the film's title.[2] To his regret, Ealing Studios rejected the commentary that he wrote on the factual sequences, preferring to use a verse-gloss by Louis MacNeice and he was also disappointed to discover that his contribution received no screen credit. But he enjoyed much of the work nonetheless and it was through his involvement with the film that he met working boater, George Smith, who appeared in the film as an extra. Generations of the Smith family had lived on the canals and George took up canal work as a matter of course. At the time of the film's production he was about to marry Sonia South, an actress by training who had undertaken war work installing electrical wiring in bombers at the Hoover Factory at Perivale before seizing the chance to volunteer on the waterways. If they came from different backgrounds – George made his way through life without much literacy while Sonia was decidedly bookish – the demands of their work drew them together and they had not known one another long before they got engaged. Meeting Sonia at the film's premier, her

New Statesman tucked under her arm, Tom thought her dismayingly earnest – 'a rather frightening left-wing blue stocking'.[3]

Among the letters to reach him from *Narrow Boat*'s fans was a missive from one Robert Aickman. Having introduced his wife Ray and himself as literary agents and signalled their mutual appreciation of Tom's writing, in his letter, Aickman went on to express his concern for the parlous condition of the inland waterways. Musingly, he proposed formation of a pressure group to take up the canal cause, remarking that 'a disinterested body of enthusiasts (but not fanatics)' could do much to improve matters. Of more immediate concern to him was the idea that Rolt and he, together with their respective wives, should meet one another in person for some discussion.[4]

The upshot was that in August 1945, the Aickmans travelled down from London to meet the Rolts at Tardebigge. With them, they brought their friends Howard and Joan Coster, both photographers who, at the time, were beginning to think about retirement and liked the idea of buying a boat and living on the canals. It was evidently a charmed encounter. Writing to Joan Coster afterwards, Ray Aickman said that how much Robert and she had enjoyed meeting the Rolts and looking over *Cressy,* and she remarked appreciatively on the peace of the place.[5] Tom and Robert had discovered that they shared an interest in the supernatural and while Tom had had some ghost stories published, Robert had not only visited Borley Rectory, supposedly the 'most haunted house in England', but also met the well-known psychic Harry Price.[6] It is not known whether or not it came up in conversation, but the fact that neither man had seen active service in the war may have been another bond between them, Tom having taken up employment with Rolls Royce, the Aldbourne Foundry and the Ministry of Supply and Robert having registered as a conscientious objector.[7]

But the nature of the two men's respective interest in canals could hardly have been more different. Tom read and wrote about them, brought an engineer's appreciation to the history of their construction and cherished the workmanship of the boat builders. By August 1945, not only had he lived on a boat for some six years, but over that time he had observed the lives of working boatmen and women at close quarters. Robert's canal enthusiasm was altogether more whimsical. As a child, he had walked with his parents near the Grand Union Canal as it crossed Cassiobury Park near Watford. Although he remembered his father's lifting him up to watch boats pass through the lock, little fondness attended his recollections. He had not liked the sensation of 'being held out over the water' and found the grimy, taciturn boat families busy with the lock paddles rather alarming.[8] Visiting Stratford-upon-Avon in the 1930s, he had taken a rather unsatisfactory walk beside the Stratford-upon-Avon canal, dismayed by its decay. It was overgrown; there was little water in it; the locks were nearly derelict, their gates 'either flapping or jammed', and the Great Western Railway Company, the canal's owners – had put up cast iron signs bearing 'threatening notices [...] directed against trespassers'. At the time, it made for a bleak experience. Once more at home, Robert read about the development of Britain's canals and the circumstances by which they came into railway company ownership but pursued the matter no further until – 'years later' – *Narrow Boat* awoke his memories and gave him the idea of campaigning for 'a new waterways world'.[9] From time to time he would think about buying a boat and often solicited Tom's advice on the subject. He was, for a while, owner of the narrow boat *Phosphorus* and enjoyed planning the layout of the vessel's interior, but his zest soon waned. Having decided that he was 'not a messer about in boats' after all, he sold *Phosphorus* to pioneering canal enthusiasts Alan and Cécile Dorward, expressing great relief at having escaped such chores as fine-tuning the engine and inspecting the hull for signs of cold-weather damage every spring.[10]

By this time, Tom was exchanging letters with fellow canal-enthusiast and emergent canal historian Charles Hadfield and was quick to bring the putative canal campaigning group to Hadfield's attention. Hadfield responded by sketching a broad perspective of what he took to be the new organisation's most pressing objectives. First, it should aspire to 'see as many navigations as possible used for purposes of trade'; second, he proposed that those waterways which were 'not useful for trade' be kept open 'for pleasure use'; third, that 'waterway amenities' – most strikingly, the right of the public to walk on canal towpaths – be preserved; and lastly, that it would be helpful to amass a 'body of knowledge about waterways', in effect, a repository of information for anyone with any sort of interest in them, be it for trade and transport, for leisure boating, or in connection with their history. To fulfil these aims, Hadfield foresaw that the organisation needed someone – ideally a lawyer – to act as its advocate with the Ministry of Transport and to be ready to give evidence at public enquiries. The new body would also need to make connections with existing waterway-related associations, both in the trade and leisure spheres, and – an area in which, since he worked at the time for Oxford University Press, thought that he might be able to assist – to build up 'a library, an enquiry service and a magazine'.[11]

An informal gathering intended to gauge the initial level of interest in the cause took place in mid-February 1946 at Robert and Ray's flat at 11 Gower Street in Bloomsbury.[12] Hadfield attended bringing with him his friend and publishing colleague Frank Eyre, a poet and canoeing enthusiast with whom he had collaborated to write *The Fire Service Today* and *English Rivers and Canals*.[13] Robert meanwhile had decided that if the canal campaign was to be successful it needed to be taken out of the remit of what he viewed as 'the small and ever more inturned [sic] group of specialist enthusiasts'. His two guests were Commander Luard, founder of a society of London sailing enthusiasts known as the Little Ships Club, and a Mr Kirkland whom he described generously if vaguely as 'an authority on all that moves'.[14] Shortly before the formal proceedings opened, Captain Richard Smith – another interested participant whom nobody remembered having invited – arrived and asked if he had reached the meeting of the Inland Waterways Association. At the time, the organisation had no name, but the Inland Waterways Association fitted its aims well enough for the founding members to adopt it.

In the course of the inaugural meeting, Robert would be elected Chairman; Hadfield, Vice-Chairman, and Tom, Honorary (unpaid) Secretary. Exhilarated, Tom returned to *Cressy* full of campaigning zeal and typed up a manifesto for the new Association, together with its rules. Since he had resourcefully kept all the fan mail that had reached him after *Narrow Boat*'s publication, he used it to draw up a list of people who might be interested in becoming members.[15] He also compiled a pamphlet for publicity entitled *The Future of the Waterways* which explained the background to the Association's formation and set out its objectives.

The post-war Labour Government proposed to take the waterways network into public ownership and while it was extensive, it was in a sorry state. The independent railway companies had sought to quell any competition from the older transport system by buying it up. Legislation required them to maintain it in good order but, as Tom's pamphlet explained, the provisions lent themselves to a certain latitude of interpretation. Although the canals' railway owners usually ensured that locks and other visible waterway features remained in good order, they were apt to neglect the essential but expensive work of dredging. For boats transporting a 'full-pay load', it meant that certain canals became so shallow as to be impassable. Marking the inevitable decline in traffic, the railway company might well apply for an abandonment order, which – should it be granted – would end their maintenance obligations at a stroke.

Yet in commercial and economic terms, water transport offered many advantages. It might be slow, but that did not matter where heavy non-perishable goods were concerned. From driving between different West Midlands' factories in the course of his work for the Ministry of Supply, Tom had had ample chance to observe the condition of the roads and recognised how vulnerable both their surfaces and the nearby buildings were to damage from heavy vehicles. With an end to petrol rationing in sight, congestion and its attendant problems could only increase. Waterways might not entirely solve the problem but at least they offered some remedy. Thinking of the friendliness with which Angela and he regularly met with from the working boatmen whom they encountered on the canals, in his pamphlet Tom described their way of life and its history in fond terms, illustrating his account with a fine photograph taken, probably by Angela, of Alfred Hone aboard his boat *Cylgate*. Where a waterway had little or no potential for commercial usage, Tom was ready to concede, albeit cautiously, that it offered the chance of development for leisure use which, after all, was vastly preferable to an abandonment order. The IWA, he indicated, would be an advocate for the waterways, seeking to remind the public both of their potential importance for transport and of the advantages that they offered for recreational enjoyment.[16]

Although the energy with which Robert took up the canal cause would find its chief outlet in campaigning – writing letters, chairing committees, addressing meetings and organising events designed to bring the parlous condition of the waterway to public attention – in April 1946, Ray and he joined the Rolts for a canal trip. *Cressy's* hull was showing signs of rot and Tom and Angela were on their way to Banbury for the Tooleys to attend to it. Whether, incidentally, either they or Tom recognised the seriousness of the condition is uncertain but, although the plan never came to anything, Mackersey notes that by this time he was already considering the idea of buying a more compact boat, suitable for travel over the northern canals with their short locks and in the Fens.[17] Except for the minor difficulties of disentangling a rope which twisted itself around *Cressy's* propeller at the Minworth Lock flight and a rising wind on the Coventry Canal that drove her towards the banks, the journey with the Aickmans was untroubled and pleasurable. Years later, in his posthumously published autobiography *The River Runs Uphill*, Robert – admittedly telescoping the nine or so months that had passed since his initial meeting with the Rolts into 'a few weeks' – recalled it with zest.[18] He saw working boats at first hand at the Aston and Farmer's Bridge Locks in Birmingham; had the thrill of spending the night in an 'eldritch wooded cutting' near Selly Oak; admired the internal decoration of 'The Strug' – strictly speaking, 'The Struggler'— the Banbury boatman's pub and met veteran boat builder and boat painter Herbert Tooley. While Tom remarked in the *Cressy* log that it had been 'a most successful journey in excellent weather', for Robert and Ray it had been an intensive close-quarters induction into canal life and lore.[19]

Robert Hale's publication of *Worcestershire* had been so badly delayed that by this time Tom had begun to think that his work on it would prove to be 'labour in vain'. Discouraged, he decided to consult Robert in his 'professional capacity' about his next book and, in consequence became a client of the Richard Marsh literary agency which Robert and Ray, working under her maiden name of Edith Ray Gregorson, co-managed which bore the (assumed) name of Robert's grandfather.[20] Born Richard Bernard Heldmann, a successful and popular novelist, dramatist and writer of short stories, he had adopted the pseudonym of 'Richard Marsh' after his 1884 conviction for forgery.[21] Since Robert's and Tom's early correspondence – Robert's 'fan' letter about *Narrow Boat* and Tom's courteous reply – had evinced such a spirit of friendship and the mutual goodwill, for Tom to become Robert's client and agree that Richard Marsh Ltd should promote his books and develop his literary career may have seemed a logical, even a propitious, step to take. But it would blur the boundaries between their friendship on the one hand, and their business relationship on the other, which may well have contributed to their future friction.

Tom toyed with the idea of writing a book about travel journals for which he had proposed the title of 'Georgian Journeys', but having solicited Robert's views on its likely commercial prospects, he abandoned it and spent much of the following few months writing ghost stories.[22] At their most distinctive, his supernatural tales concern characters – miners, perhaps, or railwaymen or other figures employed within an industrial setting – whose activity disturbs an occult power which has hitherto rested dormant within the Roltean landscape. The premise on which many of them rest finds expression – surprising as it may seem – in the context of his rigorously factual history of railway accidents and railway safety, *Red for Danger* (1955). Remarking on the sheer number of accidents that occurred on the Settle–Carlisle line, the thrusting Midland Railway Company's 1870s route through the Cumbrian fells, he casually observes, that it was 'as though the solitude resented the [...] arrogant invasion'.[23] Already, the theme of humanity's 'arrogant invasion' of earth's hidden places, often locations which possessed some forgotten primeval significance, had come to preoccupy his imagination, giving rise to a type of fiction that Robert and he came to think of as 'industrial macabre'.

Tom and Angela arranged to spend much of the summer of 1946 exploring the Irish canals. It was both a holiday and a chance for Tom to gather material for *Green and Silver*, his account of the Irish canals. By mid-June 1946, the Rolts were in Athlone, having chartered *Le Coq*, the vessel in which they undertook their voyage, from a Mr Beahan. Writing to Robert, Tom described Beahan as 'supremely vague and an incurable optimist', adding that his boat 'perfectly' reflected both characteristics.[24] In the same letter, he asked whether Robert had a stamp of his signature for use in IWA correspondence; while he could sign circulars and other Association communications by hand, sending them to him in Ireland for signature and for despatch on to the members would involve expense and delay 'which' he observed 'we really cannot afford'. Robert took the point, but it did not stop him from bombarding Tom with concerns and complaints about the inefficiencies of the IWA's day-to-day organisation.

Charles Hadfield had recently recruited a young woman named Daphne Miller to run the Association office.[25] It soon emerged that she not only lacked all command of spelling and punctuation but had very little typing skill. 'She definitely said she could type and do secretarial work', observed Tom unsympathetically from Mullingar when her shortcomings came to light, 'so it would seem that she has been guilty of misrepresentation'.[26] It was an embarrassing situation, especially since Miss Miller was a relative of Hadfield's wife, but Robert ruthlessly dispensed with her services and installed his lover, novelist Elizabeth Jane Howard known as Jane in her place.[27] It was an inspired stroke. Jane, who had recently left her husband, war hero and founder of the

British Wildfowl Trust Peter Scott, ran the IWA office most efficiently on a modest salary of two pounds ten shillings a week.[28]

But a more serious upheaval within the IWA occurred in August 1946 when Hadfield suddenly resigned from the post of vice-chairman. In time, he confided his reasons to Tom who replied in a letter that revealed much about the cross currents at work in the Association's Council. 'I am sorry that that you could not see eye to eye with Aickman', wrote Tom:

> and [I] feel that if it had not been for my absence in Ireland, I might have been able, by suitable diplomacy, to have kept a balance of power [...] Aickman has not, I know, an easily likeable personality, and is apt, on this account to antagonise people....[29]

From these remarks, it sounds as though he too had begun to wonder how long he and Aickman could continue to work together harmoniously. To Aickman himself, Tom said outright that he thought Hadfield's departure was 'a pity.' 'The council seems to be dwindling to a dictatorship' he added. 'Who remains? Cannot we, out of our membership, elect a few useful people?'[30]

Before they departed on their Irish trip, Angela and Tom had moved *Cressy* to Banbury boatyard with the intention that Herbert Tooley should attend to her hull but somehow he got badly behind schedule and when the Rolts returned, they found their boat untouched at her mooring.[31] They had promised to take Mima on a nostalgic visit to Llanthony, but as soon they reasonably could, they returned to Banbury to redecorate the boat's interior and – since finding suitable temporary accommodation in the town proved impossible – hired a caravan in which to live while Mr Tooley caught up with himself. Painting and assisting with the other works in progress on *Cressy* while at the same time composing ghost stories for the collection that would become known as *Sleep No More* took all Tom's attention, but a welcome diversion came in mid-October 1946 when George and Sonia Smith came through Banbury with a consignment of goods. Remembering them from the Ealing film, Tom and Angela spent an evening on their boat *Halifax* and Sonia told them some of the ghost stories in circulation on the canals. It was just the thing to catch Tom's imagination and when she mentioned that George had learned many canal folksongs from his father, the Rolts urged her to record them. Writing to Ray, Tom suggested that the IWA might consider creating a special class of membership for working boatmen who could not be expected to pay the standard guinea subscription.[32]

As autumn advanced, the weather grew colder; by the end of October, longing to get out of the caravan, Tom appears unofficially to have joined Mr Tooley's workforce, engaged on *Cressy*'s overhaul. His efforts paid off. By November 1946, Angela and he had once more ensconced themselves aboard their boat. While Tom worked on the draft of *Green and Silver*, Angela negotiated with

Blinkhorns, the Banbury Photography business, to print her photographs for use in IWA material.[33] Both Rolts, in other words, were extremely busy, but Robert in his agent's role insisted that Tom should also prepare a script – a substantial forty-five minutes in length – for broadcast by the BBC on the subject of the waterways.[34] Admittedly it carried a fifty guinea fee, but the work took hours that Tom could not easily spare. Robert, with his modest private income to boost his agency earnings, could afford to devote considerable time and energy to the cause. Not all the IWA council members were so advantageously placed.

For the duration of the Second World War, the British government had controlled the nation's docks and railways. The post-war Labour Government proposed to take the railways and associated docks into public ownership and, to this end, established the British Transport Commission. Collectively, it had been rather slow to recognise that, since many canals were in railway ownership they would, perforce, also be nationalised. Tom drafted a memorandum detailing the IWA's responses to this development. Among other provisions, it stipulated that 'existing navigations should be kept in first class order both for commercial and pleasure traffic'; that a 'systematic policy of dredging should be carried out which should aim to achieve a minimum depth of five feet on all waterways'; and that 'steps be taken vigorously to promote the use of waterways for recreation'.[35] Robert despatched a copy to Minister of Transport, Alfred Barnes MP and, in consequence, the IWA was invited to send a deputation to meet the Minister and consider the potential consequences of the Transport Bill for the inland waterway network.[36]

The meeting took place on 5 March 1947. Representing the IWA were the Association's newly appointed president, Sir Alan Herbert, MP, otherwise known as A.P. Herbert and author of *The Water Gipsies* (1930) and *Holy Deadlock* (1934) among other works; Robert and Tom, Chairman and Honorary Secretary; council member Captain Richard Smith and working boatwoman Sonia Smith. Once the president had introduced his colleagues, Robert explained that the Association's chief objective was the restoration of the waterway network to efficient order. Not only, he suggested, were there good reasons, firstly, for increasing usage of the waterways for commercial purposes but, secondly, they were also a potentially valuable leisure amenity. In recent times, he asserted, as a consequences of neglect, many of them had either been abandoned or had fallen into dereliction. Both the Kennet and Avon and the Stratford-upon-Avon canals were, to all intent, unnavigable. Even if a waterway did not lend itself to commercial use, it could be maintained for leisure purposes at relatively low cost, but the opportunities for recreation and enjoyment vanished with its abandonment. To bring an abandoned canal back into usage was a much more formidable and expensive task than keeping an existing waterway open.

These were powerful arguments, which both Tom and Captain Smith were quick to endorse. Speaking on behalf of working boaters, Sonia made the point that inadequate dredging of the canals in recent years led to a reduction in the tonnage that boats could carry. For boaters, it could mean a significant loss of income. In passing, she also made a plea for the narrow boat crews to receive the same generous food ration as the estuarial boatmen – a matter which the Minister advised her to pursue with her trade union.[37] What, if anything, the meeting achieved is hard to say. Mr Barnes expressed his hope that the IWA might be a channel of communication between the public on the one hand and the soon-to-be formed Docks and Inland Waterways Executive on the other, but declined to give any assurance that no further navigations would be closed.[38]

Even before the IWA deputation's meeting with a Minister of State, Robert wanted the organisation to undertake 'some campaigning [...] by deed instead of merely by word'.[39] The canals were not yet nationalised and he and Tom wished to draw public attention to their fate under railway ownership. To make a clear point, they opted to focus on the GWR's Stratford-upon-Avon Canal which was effectively blocked; and on the Welsh Section of the Shropshire Union Canal – the 'Welsh Canal' as Tom calls it – which the LMS had obtained leave to abandon in 1944.[40]

The blockage on the Stratford-upon-Avon Canal had come about in 1945 when a lifting bridge at Tunnel Lane, Lifford, collapsed under the weight of a heavy goods vehicle.[41] Rather than replace it with another lifting bridge, the GWR installed a fixed steel structure in its place which served the needs of road users but made the canal impassable by any vessel larger than a canoe. Recounting their experience of taking a cargo of glucose from London to Cadbury's factory at Bourneville, George and Sonia Smith had been eloquent upon the subject of the inconvenience caused to boaters. After all, the Stratford-upon-Avon Canal had promised to be the Smiths' most direct route, but when they got warning from another boat crew that they were unlikely to be able to get under the Lifford bridge, it had left them with no option but to take the long way round, going into the middle of Birmingham and then working their way out to Bourneville via Worcester Bar.[42] In March 1947 – the same month that the IWA delegation met Mr Barnes – the *Birmingham Post* published a letter from IWA Midland's Organiser, John Stables, drawing attention to the difficulties of navigating the Stratford-upon-Avon Canal.[43] Around the same time, primed by Robert, IWA Member Lord Methuen obligingly asked some pointed questions in the House of Lords about the legality of the obstruction. It resulted in an assurance from the GWR that the Lifford bridge would be lifted on notice of an intended passage at any time[44] Seizing his opportunity,

Tom decided 'to write to the GWR, not as IWA Secretary, but as a private individual and simply ask for a passage through the canal'.[45]

His chosen date was 20 May 1947. Although the exercise of his right of navigation promised to catch the public eye, Tom anticipated it with mixed feelings. 'Personally, I loathe anything of this sort', he told Robert, dismissing out of hand any proposal that the IWA might mount an 'organised rally at the famous bridge'.[46] Nevertheless, the sheer David-and-Goliath effrontery of the enterprise – 'challenging the might of the GWR' as he expressed it – clearly caught his imagination.[47] 'Lay at Kingswood troubled by reporters and IWA members' he wrote in the *Cressy* log for 18 May, anxious lest *Cressy* should either stick beneath the bridge under the eyes of the press, or get damaged in the course of her passage along the canal. The following day Robert, who had taken the Birmingham train, arrived to join him on board *Cressy* after lunch. They worked their way up the nineteen Lapworth locks, acutely aware of the presence of *Bilster*, the motorboat which the GWR had chartered to precede them through the bridge, going just ahead with a number of official personnel on board. All went well until they reached the near-derelict canal's summit, where the weeds were so thick that they had to 'resort to bow-hauling'. Wryly surveying *Bilster*'s progress and thinking that the boat's chances of getting through the bridge were slight, Tom hung back, hoping that he would not find his passage blocked by the pilot vessel's wedging herself in a bridge hole before they had even reached their destination.[48]

It was precisely what happened. Having spent the night near Earlswood Reservoir, Tom and Robert set off aboard Cressy the following morning only to find that *Bilster* had grounded in a bridge hole right in their way, jammed fast 'on a mound of debris'. With *Bilster*'s crew and passengers they tried to dislodge her, but achieved nothing until they managed to persuade the driver of a tractor who had been ploughing a nearby field to abandon his work and tow her clear of the obstruction.[49] Not long afterwards, *Bilster* grounded again but this time, with some adroit steering, Tom managed to manoeuvre *Cressy* ahead and even contrived to 'pass a line over *Bilster*'s fore-end stud', make it fast and pull the GWR boat free. That *Cressy*, now the 'trail-blazer', should have grounded soon afterwards was humiliating, but Robert and Tom at least managed to haul her clear unaided. At this stage *Bilster* and her passengers abandoned the mission and signalled to Tom to cast them off. Then, proceeding with the utmost caution and at the slowest conceivable pace, *Cressy* advanced through the King's Norton tunnel. Emerging into daylight, Tom and Robert found a crowd of spectators on hand to witness their passage under the bridge which 'a posse of overalled GWR gangers' had jacked up in anticipation of their arrival. *Cressy* just about got underneath it, albeit with much manhandling 'beneath the girders'.[50] It was hardly a triumphal progress but Robert and Tom had accomplished what they set out to achieve.[51]

If the Lifford Bridge adventure went ahead in full glare of publicity, when Tom turned his attention to the Welsh Section of the Shropshire Union Canal and resurrected his ambition to take *Cressy* over the Pontcysyllte Aqueduct, he set about it in an unassuming way. Angela, at least to begin with, was his sole companion. But at Nantwich, the Rolts met the canal-minded Grundy family – Mr and Mrs Grundy, sons Christopher and Martin and Mr Grundy's elderly if stalwart sister – holidaying aboard their cabin-cruiser *Heron*. They too had the idea of exploring the Welsh Canal and, foreseeing that it was likely to be overgrown and neglected, the two parties decided to join forces so that they could help one another if it became necessary.[52] It was a wise precaution, since – despite the care with which Tom greased *Cressy*'s sides and planed away the high spots on her hull – deformed lock chambers, silted channels and low water levels still made for a slow and difficult journey.[53] Thick weed-growth was a major obstruction and at Wrenbury, they decided to hitch *Heron* behind *Cressy* so as to make a united assault. Describing to Robert their progress between Wrenbury Bridge and Marbury Lock, Tom wrote that:

> It consisted of (1) Angela and Mr and Mrs Grundy bow hauling *Cressy*; (2) *Cressy* with self at tiller driving furiously; (3) *Heron* in tow behind *Cressy* and occupied by Grundy's elderly sister in a great state of agitation. She is nearly blind, stone-deaf and was seen to be steering with one hand and brandishing a huge tortoise-shell ear trumpet with the other.[54]

Weed that 'looked solid enough to walk on' repeatedly wound itself around the boats' propellers which meant that they rarely covered more than about two miles in a day. Surveying the canal, Tom almost despaired. 'Given even a small regular traffic', he observed writing to Robert, 'the weed trouble would soon disappear.'[55] To Hadfield, he commented that 'The murder of this canal and its present state are as shocking a story as any that I have come across'.[56]

Struggling to make any progress, Tom could not help recalling his joyous cruise along the self-same waterway some seventeen years previously with his Willans relatives. By chance, he had sent Kyrle and Hero a copy of his recently published book *High Horse Riderless* only the previous month. Why he imagined that his critique of life in 1940s Britain would appeal to them it is hard to imagine, unless he assumed that, as relations, they were likely to be curious about his publications. *High Horse Riderless*, its title drawn from the closing lines of W.B. Yeats' poem 'Coole Park and Ballylee, 1931' was a polemic. Although Tom claims that the book sprang more from his wish to 'clarify and codify' his ideas than 'with any thought of publication' it addresses its readers in the style of a campaigner rallying followers to a cause.[57] 'The narrow road to the village needs widening', it announces at one point:

> so that Bentleys and S.Ss may speed home safely; then the bridge over the millstream will be rebuilt to take the weight of the Corporation double-decker in the wake of which will follow the cinema and the chain store. The mill itself will acquire a new lease of life as a bathing lido [...]

'This', the passage concludes, 'is the process of infiltration by which our new material civilisation has engulfed the older order'.[58] The need for society to return to the 'older order' – learning to be self-sufficient and recovering its connection with the nurturing land – is the book's theme. Perhaps the most resonant of the 'Conclusions' that Tom draws from his survey lies in his statement that:

> a self-sufficient society must necessarily be based on a prosperous and populous agricultural community, and that the purpose of industry must be to serve that community and not vice versa.'[59]

Not surprisingly, Kyrle and Hero Willans had not found it at all to their taste. 'I have just had a letter from Mrs Willans', Tom soon informed Robert, 'complimenting me on the book but saying it was rather beyond her. What is beyond her is evidently so much caviar to the general where the Commander is concerned. No flowers for the late Secretary!'[60]

At Ellesmere Canal Maintenance Depot, in between 'trimming hedges and scything the grass on the canal banks' the staff warned him that a landslip beyond Chirk had blocked the canal.[61] It put paid to any question of reaching Pontcysyllte. The Grundys by this time had gone on ahead but the Rolts stayed for a few days in Ellesmere. Then Tom decided that he did not entirely care for the available moorings and before long, he and Angela retreated to Hampton Bank near Welshampton where they explored the rural branch lines of the Welsh Border.

All this time the weather was fine and warm which, while it made for a delightful holiday, did nothing to remedy the shortage of water in the canal. Already low, in the course of a four-week drought, the level fell by more than an inch. Then, it suddenly began to sink at worrying speed. From his anxious enquiries, Tom learned that a canal bank had burst near Chirk and that the feed from the Dee near Llantysilio had been shut off for repair. His bold endeavour to travel along a canal over which he had no right of navigation looked as though it was about to rebound on him, and he reflected ruefully that. '*Cressy* could sit on the mud at Hampton Bank until doomsday for all that L.M.S. railway cared'.[62]

Neither the Rolts, nor the Grundys who had recently rejoined them, had any wish to become stranded. If they were going to get away, it was crucial that there should be enough water in the canal for them to reach Hurleston Junction. Fortunately Hampton Bank had a telephone box and, taking matters into his own hands, Tom called Mr Howell, the lock-keeper at the Grindley Brook Flight beside the Hurleston Reservoir to explain the situation. Both men realised that if the bypass culvert by which the canal fed the reservoir were temporarily to be shut off, enough water would accumulate in the canal to resolve the dilemma. While Mr Howell was not unsympathetic, he was not prepared to shut off the water supplies to the reservoir 'without authority from Crewe' – that is to say, from the L.M.S.[63] At this point Tom appears to have phoned the railway's Crewe works. It is not known precisely what exchange took place, but his laconic statement in the *Cressy* log for 17 August 1947 – 'Persuaded L.M.S. Crewe to hold the water at Grindley Brook so that we could attempt return' – tells its own tale. At 6 p.m. Mr Howell confirmed that he had complied with his instructions and dropped the by-pass paddle; *Cressy* and *Heron* set off, *Cressy* scraping the bottom of the canal as far as Bettisfield. They reached Platt Lane Wharf around 9 p.m. and moored there for the night.

The following day, shortage of water continued to bedevil their journey. From his memory of IWA membership lists, Tom recalled that an IWA member – one Squadron Leader Gerald Edward Livock DFC, AFC – lived in the area, and contacted him for assistance. Happy to help, Livock and his wife drove to Platt Lane Wharf where they helped to remove all the ballast and fuel which had been in *Cressy*'s aft hold and transferred it – together, presumably, with *Heron*'s cargo – to their car, whereupon Mrs Livock drove off, complete with a substantial load, to Nantwich. The boating party decided that *Cressy*, now riding as high as the limited water allowed, should tow *Heron*, with Angela, Christopher Grundy and Squadron Leader Livock bow-hauling from the towpath.

Their progress was far from easy and *Cressy* grounded on the scour near Blackoe Cottages, but with some help from the cottagers, they reached Nantwich at 6 p.m. on 20th August. There, having restored the boats' respective contents, the Rolts, the Grundys and the Livocks shared a celebration meal at the Crown Hotel.[64] For all its frustrations, the journey had furnished Tom with ample ammunition for his canal crusade.

Above: The wedding of Lionel 'Lily' Rolt to Jemima 'Mima' Clarke-Timperley, 25 September 1906. (Courtesy of Richard and Tim Rolt)

Right: Tom with his beloved Welsh nurse, Mary Gwynne, at Cusop. (Courtesy of Richard and Tim Rolt)

Lily and Tom with the Williamson combination bike that his father bought 'for family use'. (Courtesy of Richard and Tim Rolt)

Above: Pitchill, c.1927. Tom's photograph shows works foreman, Percy Lester, on the ploughing engine which Douglas Bomford had designed and constructed from Sentinel and Fowler parts. (Courtesy of Richard and Tim Rolt)

Left: Tom as a Cheltenham College boy, wearing formal clothes and a mortarboard, c.1922. (Courtesy of Richard and Tim Rolt)

Kerr Stuart, California Works, Stoke-on-Trent in the early 1920s. Two boilers – one narrow gauge and one mainline – are on the North Staffs Railway wagons. (Courtesy of Vaughan Pomeroy whose grandfather – Kerr Stuart's shipping clerk – stands on the right)

Above: Kerr Stuart locomotive hoisted aboard the cargo ship *Beljeanne*, bound for Buenos Aires, 1920. (Courtesy of Vaughan Pomeroy)

Below: Tom in the Phoenix Special at Prescott, 16 May 1938. (Courtesy of Richard and Tim Rolt)

Above: VSCC Speed Trial, Bramshill House, 1936. Tom drives a Flying Fifteen Darracq with passenger Aubrey Birks. (Courtesy of Richard and Tim Rolt)

Below: Tom and Angela in Tom's 1902/3 Humber, registration number FH 12, in 1937–8. (Courtesy of Richard and Tim Rolt)

Tom and Angela Rolt aboard *Cressy* on their honeymoon, *Leicester Daily Mercury* 12 August 1939. (Courtesy of *Leicester Mercury* and Reach PLC)

Above: *Cressy* at her Tardebigge mooring, the Rolts' home for nearly five years. (Courtesy of Richard and Tim Rolt)

Right: 'No Trains Today'. Tom at Pendre on a wet day in 1943 making an early and not entirely satisfactory visit to the TR. (Courtesy of Richard and Tim Rolt)

Above: Campaigning, May 1947 – *Cressy* comes under the Lifford Bridge on the Stratford-upon-Avon Canal. (Courtesy of Richard and Tim Rolt)

Below left: *Leicester Daily Mercury* 15 August 1950: Angela swabbing *Cressy*'s deck at Market Harborough. The newspaper caption reads: 'The striped skirt in which Mrs Rolt does her boat chores is made from pillow ticking – easily washed, ironed and not too flimsy for the waterside'. (Courtesy of the *Leicester Mercury* and Reach PLC)

Below right: Sonia in the 1940s. (Courtesy of Richard and Tim Rolt)

Above: Tom at Towyn Wharf Station, early 1950s. (Courtesy of Richard and Tim Rolt)

Below: Tom in the Alvis at Bryneglwys, early 1950s. (Courtesy of Richard and Tim Rolt)

Figurehead of the brigantine *Emily Burnyeat*. A Mr Keener – sometime a ship's rigger – showed Tom the figurehead when he was in Whitehaven researching *Mariners' Market* in the 1960s. (© Charles Miller Auctions, 2016)

A rare picture of Tom and Sonia from the 1960s. Severn Tunnel Pumping Station, Sudbrook. (Courtesy of Richard and Tim Rolt)

Chapter Nine

After Lifford Bridge

To foster interest in waterways' history and traditions, Tom thought that the IWA might put on an exhibition devoted to canal-related arts and artefacts. He raised the question of staging it at the Mansard Gallery of Heal's rather splendid shop on Tottenham Court Road with his sometime VSCC acquaintance, Anthony Heal, who viewed the idea with enthusiasm.[1] Collecting material for display was a considerable challenge, but Hero Willans offered her 'boat-women's' sunbonnets'; Charles Hadfield, his early canal maps and Tom was confident that through his contacts among boaters, with the Tooley brothers, with Frank Nurser the Braunston-based boat-builder and painter, and with historically-minded canal engineer Charles Hadlow, he would be able to source dippers, cans and pitchers, horse brasses, windlasses, tiller bars, ornamental ropework and similar items. Angela's photographs of canal life would make a striking contribution to the event. In the course of their return from Hampton Bank, Tom decided that instead of mooring at Banbury in their usual way, Angela and he would moor instead at Gayton on the Grand Union Canal which was a convenient place from which to collect likely material for display. They managed to borrow examples of boaters' traditional garb, including the traditional corduroy jackets and trousers of the men, together with their plaited or embroidered belts and the tiered skirts and lace shawls once worn by boat women. Overall, the assemblage offered an insight into the working lives of what, to many people, may well have been an almost invisible sector of the population and, what was more, illustrated the traditions and skills that the people who worked and lived on the canals had built up over the years.

IWA president Sir Alan Herbert opened the exhibition on 14 October 1947 and Robert introduced him to the invited guests with a fulsome observation upon Herbert's reputation for giving 'glamorous parties on the Grand Union'. Less than amused, Herbert replied to the effect that he had no idea what Robert meant since the only person with whom he travelled on the canals was his wife.[2] If Robert's tactlessness made for a less than propitious ceremony, the exhibition proved immensely popular and would soon travel to different venues round the country. From Birkenhead where it opened in January 1948, it went on to Luton, Bristol, Darlington, Leeds, Northampton and Leamington Spa.[3]

Tom and Angela, who had sourced the exhibits and had direct knowledge of their ownership and provenance, made themselves responsible for ensuring that they travelled safely and that the touring show received adequate publicity. If it created much work, it also brought the waterways' cause and, by extension, the IWA, to public attention. What was more, it led to Tom's receiving an attractive commission. 'My publisher [Philip Unwin] went to the exhibition', he confided to Hadfield, 'and heard people asking questions like "How does the water get into the canal?" etc. and is very keen for me to do a general popular book about canals and canal working. [...] I think this is rather a good notion.[4]' The book, entitled *The Inland Waterways of England*, would take took shape over the following two years.

For all its good results, the exhibition failed to cement amicable relations within the IWA Council; strikingly, from 1948 onwards the tensions between Robert and Tom deepened. IWA business, Tom came to think, had taken over a large part of his life. Not only did he resent the constant 'need to be at a certain place by a certain date', but he found that being 'out of reach of the post for more than a day or two at a time' – something that his peripatetic way of life made it difficult to avoid – caused him and his colleagues much anxiety. 'It is ironical', he wrote to Robert in June 1948, 'that I, who adopted this way of life' – that is to say, living aboard a boat on the waterways and writing for his living – 'to be rid of such things should have saddled myself with them of my own volition.'[5] The sheer volume of his IWA-related correspondence bears witness to the amount of time that he spent writing letters. It was not what he had anticipated when he conceived his design for living.

There were other annoyances, often in themselves niggling, but at the same time symptomatic of a certain lack of openness. In October 1947, remarking with palpable embarrassment that he did not care to concern himself 'with the private affairs' of other people, Tom took Robert roundly to task for allowing him to believe that Jane had been 'working away as usual in London when in fact she was doing nothing of the kind'.[6] Robert and Jane, he gleaned, had been away together and while he raised no question concerning the propriety of the arrangement – 'On the question of how Jane spent her canal holiday and who with, I could not care less', he insisted – but the fact he had been misled damaged what he termed the 'mutual trust' among colleagues on which the IWA relied.[7] Robert admitted that Jane and he had hired a boat for a trip on the Thames, and reminded Tom that she worked hard for the Association and needed a holiday, but Tom's sense that a kind of cabal was at work from which Angela and he were excluded lingered on.[8] Writing to Hadfield long after the event he would complain about the 'distasteful way in which the Aickmans and their particular little clique have always tried to use the IWA as a vehicle for self-aggrandisement'.[9]

After Lifford Bridge

Angela and Tom had hoped to revisit Ireland in the summer of 1948, but Robert committed them to joining Jane and himself with Anthea and James Sutherland on another campaigning trip designed to publicise the deteriorating state of the waterways. It was to be a six-week exploration of the north of England canals. The plan was that they would travel from Stone in Staffordshire to Ashton-under-Lyne by the Trent and Mersey Canal, Macclesfield, Peak Forest and Ashton Canals; cross the Pennines by the Huddersfield Narrow Canal; head towards Leeds by the Calder and Hebble and Aire and Calder Navigations; follow the Leeds and Liverpool Canal back over the Pennines to Leigh via Wigan; and from Leigh take the Bridgewater Canal to Preston Brook where it joined the Trent and Mersey Canal, and thence back to Stone. Formidable in terms of distance, the fact that many of the canals were, at the time, in a shocking state added to the rigours of the journey. Because *Cressy*, at seventy feet in length, was too long to fit the northern locks, Tom and Robert approached IWA member Rendel Wyatt, who had recently set up the Canal Cruising Company and chartered *Ailsa Craig,* a converted lifeboat, for the trip.[10] Tom observed that her hull was 'very rotten' and in the course of the journey he often had to address various troubles with the engine.[11]

Although she was compact enough to cope with the route, *Ailsa Craig* was too small to accommodate a party of six in any comfort. Collectively, the party agreed that the Sutherlands, Robert and Jane would start from Stone; Tom and Angela would travel by rail to Diggle and come aboard at the south-western end of the Standedge Tunnel – Britain's longest canal tunnel on the highest summit level. For its transit, all six of them would be on the boat. Then the Sutherlands would disembark at Marsden at the tunnel's north-eastern end and the Rolts would return with Robert and Jane to Stone. In view of the limited available space on *Ailsa Craig*, Robert purchased a tent to provide extra sleeping accommodation and, rather to Tom's irritation, debited its cost to the IWA.[12]

Tom and Angela made a 'leisurely journey' to Birmingham, having detoured via the Grand Union Canal to Braunston where Frank Nurser brought all his artistry to bear upon renewing *Cressy*'s interior decoration. By slow stages, they joined the Severn at Tewkesbury and from Gloucester, followed the Gloucester and Berkeley Ship Canal to Sharpness where they found what Tom considered a 'perfect mooring' well out of the way of the commercial traffic. There they stayed for a month, enjoying the expansive views over the wide Severn, watching the shipping, and exploring the locks and docks. They visited the immense Severn railway bridge and Tom scaled the iron ladder leading up to the cabin from which a signal-man-cum-engineman regulated traffic on the line while minding the mechanism which opened the bridge's swing-span. The Sharpness pubs were depressingly dingy, but the inn up river at Purton was welcoming and popular with bargemen and netters of salmon.

In the course of their stay, Tom met Brian Waters, author *of Severn Tide* which Angela had given him the previous Christmas; Bill Willans who had recently been de-mobbed also paid them a visit. Their time in Sharpness had nothing to do with the IWA and its campaigns, and when the time came for the Rolts to head north towards Stone, they went on their way with some regret.[13]

Since the LMS, sometime owner of the Huddersfield Narrow Canal, had officially abandoned it in 1944, in order to travel over it the party required the consent of the British Transport Commission. Most obligingly, the commissioners not only gave permission for the journey but also promised that a work gang would be on hand to help. It was just as well because, unknown to the Rolts, the trip had almost ended ignominiously soon after it began. Venturing below deck as they passed through a lock on the outskirts of Ashton-under-Lyne, Anthea Sutherland found the cabin floor awash and at once guessed what must have happened. A sharp object on the canal bed had holed *Ailsa Craig*'s hull. Everything that they had brought for the journey – notably Robert's poetry books and silk dressing gowns – required unloading. While the rain poured down, the work gang pulled the boat out of the lock and into the next pound which, thoughtfully, they drained before quitting the scene. James Sutherland, an engineer, directed them to shore up the boat with some of the scrapped aluminium aircraft parts which littered the waste ground beside the canal. Predictably, the exercise caught the attention of the local people who were quick to offer food and shelter, which the travellers, concerned about the consequences of even temporarily abandoning the unsecured boat, declined. Since his typewriter was unscathed, Robert dictated to Anthea a letter to the Chairman of the Docks and Inland Waterways Executive with the result that two Mersey shipwrights soon arrived to repair the damaged hull, together with canal foreman Wilfred Donkersley, who stayed with them all the way to Huddersfield. While they were stranded, they caught much local curiosity, allowed themselves to be interviewed in the local paper and accepted the offer of seats for a show at the Ashton Music Hall where the performers made much of their plight.[14]

Unaware of the disaster, Tom and Angela waited at Stone until they received their 'signal' to join *Ailsa Craig*. Only then did they take the train to Diggle where they joined *Ailsa Craig*'s crew. Whatever misgivings Tom may have harboured about his companions, the tunnel itself caught his imagination to such a degree that he described passing through it as 'one of the most eerie and sensational experiences' that he had ever known. Because the Pennine Rock through which the canal company had driven it was unusually hard, for much of the tunnel's three and half mile length, it did not require to be lined with brick. As a result, much of it resembled a cavern rather than a man-made thoroughfare, and variations in the tunnel's diameter and the absence of any towpath both served to heighten the resemblance. The fact that cross-galleries linked the canal tunnel (completed in 1811) to the Huddersfield and Manchester Railway Tunnel (completed in 1848) meant that 'thunderous reverberations' and blasts of smoke from passing trains heightened the sinister aspect of the place.

In the course of their transit which, Tom recalls, took 'at least two hours', at one point *Ailsa Craig* stuck fast under the low vault. Recollections as to how they freed themselves differ. In *Landscape with Canals* Tom reflects that having six people on board at least meant that the boat rode relatively low in the water besides providing plenty of hands to assist in pushing her clear of the constricting walls. James Sutherland, with a slight but significant difference of slant, remembers Tom's suggesting the bold strategy of reversing and then 'taking a run' at the narrows. 'There was', he recalls, 'a worrying moment with scraping of paint and some splitting of wood as the boat pushed down in the water' and 'Then suddenly we were through.' Long after the event, Sutherland remembered that it was during the tunnel's transit that Tom's experience became fully apparent. Steering with vast skill, he deftly navigated 'jagged rock surfaces with varying clearances' to find *Ailsa Craig* a safe passage 'through what seemed like great caverns where the rock had fallen during construction'.[15] Robert, in his account of the passage, makes much of the jovial presence of Wilf Donkersley and his companions who apparently preceded *Ailsa Craig* through the tunnel in a narrow boat, Mr Donkersley regaling them 'with a sequence of songs' along the way.[16]

No-one cared for sleeping in the tent. The night that Tom and Angela spent in it, a herd of young bullocks repeatedly blundered into the guy-ropes, leaving Tom to scare them off, retrieve the uprooted tent pages and pitch it time after time in the dark and teeming rain.[17] Jane had bitter recollections of pitching it in bad weather. Even worse than the discomfort of wet nights spent under canvas was the acrimony that Robert brought to the whole adventure. Jane, always an acute observer of relationships, recognised his capacity to poison the happiness of people around him all too clearly. Not only did he expect 'everything to be done for him', but 'the only thing he wanted to do in a boat was to steer it' and, frustrated to discover that his steering skills were extremely limited, he turned irritable. Not only did he argue with Jane and pick quarrels with Tom, of whom he was growing increasingly jealous, but he even embarked on a 'flirtation' with Angela which, Jane succinctly remarks, 'didn't go down well with anyone'.[18] Doggedly, they continued on their way through Huddersfield, Wakefield, and then over the filthy waters of the Aire and Calder Navigation as it approached Leeds. Having joined the Leeds and Liverpool Canal, they ascended the Bingley five-rise to traverse seventeen level miles over the Yorkshire countryside. Once they had passed through the Foulridge Tunnel, they came down into Lancashire, to skirt Burnley on Robert Whitworth's vast embankment. Before long, they joined the Bridgewater Canal which gave Tom the chance to explore Worsley Basin before they made their way to the junction with the Trent and Mersey Canal at Preston Brook and thence back by slow stages to Stone.[19] In *Landscape with Canals*, Tom says little about this part of the journey, which may suggest – it is no more than conjecture – that he did not entirely care to remember it.

Late in 1948, in a rather surprising development, the Basingstoke Canal came up for sale at auction. Constructed in the late eighteenth century to connect Basingstoke with the River Wey and thence to the Thames, it had for years been in the ownership of the Harmsworth family. When Alec Harmsworth died in 1947 his sons decided to sell it to the highest bidder.[20] The IWA might have wished to buy it, if only for the sake of showing the Waterways and Docks Executive how a canal ought to be managed, but its constitution did not allow it to enter into business – which would have included managing a waterway – and, anyway, its funds were wholly inadequate for the purchase.[21] IWA members' collective interest in the Basingstoke Canal's fate was nonetheless great and one of their number - Lewis Edwards, known as Teddy - convened a meeting in Woking on 11 December 1948 to discuss it. Robert attended and from those audience members who asked intelligent questions or made sensible observations, he formed an *ad hoc* Basingstoke Canal Committee. While this committee had no formal connection with the IWA, it afforded the IWA Council and committee a means of keeping in contact with developments at Basingstoke.

Over the following months, the IWA received not only much correspondence about the Basingstoke Canal but also donations totalling about £3000. At the auction, which took place in Aldershot on 1 March 1949, the only bidders were Basingstoke Canal Committee member, Joan Marshall, and local building contractor, James Chuter who, having made an initial bid of £3000 withdrew soon afterwards.[22] Mrs Marshall acquired the canal for £6000 and went on to bid an additional £3000 for ancillary lots which included a lock cottage.[23] Several newspapers enthusiastically reported that she had been acting on the IWA's behalf.[24] The London-based *Daily News* even imagined Tom on 'his painted barge somewhere in the Midlands' cheering over the auction's outcome.[25]

Nothing, in fact, could be further from the truth. The day after the auction, Mrs Marshall telephoned Robert to tell him that he must not think that the IWA had acquired the canal. Members of Robert's Basingstoke Canal Committee had, it emerged, taken stock among themselves on the evening before the auction. They had indeed decided to set up a new committee – the 'Basingstoke Canal Purchase Committee' – for the purpose of acquiring the canal but they had no intention of being in any sense bound by IWA objectives. Instead, having constituted themselves as a self-governing body, they reckoned to manage their acquisition on their own terms.

Tom, who watched the Basingstoke saga unfold from *Cressy*'s Banbury mooring, admitted frankly that he had taken 'the gloomiest view' of the enterprise all along and 'wouldn't be surprised' by anything that Robert told him about it. Were the newly constituted Basingstoke Canal Purchase Committee to retain charge of the waterway, it could cause much future trouble. 'This is where my uncle Kyrle will jump in with both feet', he wrote, adding ominously that he knew 'from experience whither uncle's schemes always lead.'[26]

The way that events unfolded bore out his misgivings. Cyril Styring, an IWA member who had pledged a large contribution to the canal's purchase, withdrew it, unhappy with the direction that the canal purchase had taken. A meeting arranged with the intention of bringing the IWA and the Basingstoke Canal Purchase Committee together in pursuit of a common purpose achieved nothing. When Tom, who may have remembered the Basingstoke Canal both from his days at Hartley Wintney and his brief employment at Thornycrofts, visited it at the end of March, he found it in a shocking state with 'much weed, little water and lock gates in a poor way;' he expressed the hope that he had 'picked a particularly bad spot'.[27] But as Robert and Tom soon realised, the canal's new owners showed little intention of improving it. Apparently, when the IWA drew Mrs Marshall's attention to its parlous condition, she explained that its maintenance lay in the hands of various unknown outsiders who had financed its purchase but who 'could not find the money' to undertake the measures that the Association wished to promote.[28]

Kyrle Willans – himself, an IWA member – had fond childhood memories of the Basingstoke Canal; he liked the idea of seeing 'the pounds full, the locks working and boats of every kind using it' and to achieve this end, he recommended installation of a pumping plant.[29] Dismayed by what he saw happening to the waterway, he bombarded Robert with abusive correspondence.[30] Early in July 1949, Robert wrote to Tom who was in North Wales at the time, apologising for troubling him, but concerned that Willans was 'simply out to injure' the IWA and to cause as much trouble as he could. From his letters, Robert evidently thought that Willans, who appeared to have had lost all mental balance, was trying to blackmail him. Wishing to see him removed from the IWA, Robert asked Tom whether there was anything that he could do 'through family channels' to ease matters.[31]

In his reply, Tom conceded that his uncle had 'touched a new "low" in imbecility and pettiness'.[32] He advised Robert to reply to the effect that although the IWA Council – whom Willans refused to meet – were ready to 'elucidate' to him the circumstances that surrounded the sale of the Basingstoke Canal, the issue did not lend itself to explanation in correspondence. From Willans' threatening to call 'a special general meeting to discuss the Basingstoke question,' it sounds rather as if he may have been hoping for an opportunity to discredit both Robert and Tom in front of the IWA membership.

A few days later, Robert wrote again with further insulting enclosures from Willans. Exasperated, Tom had nothing to add to his earlier advice. Mima, by chance, had recently met Hero in Bath and thought that she had aged perceptibly. 'Poor lady' observed Tom, telling Robert about the incident, 'One is scarcely surprised'. In the same letter, he remarked that 'Willans' lunacy' had become increasingly pronounced, 'which' he added, 'is all to the good since it seems to involve his falling out with everybody very rapidly'.[33]

It was a flippant observation, intended to shed the light of bracing humour upon the intractable situation that beset the IWA Council. But Tom mistook his man. By now, Robert was in no mood for humour. Instead, he was spoiling for a fight.

Chapter Ten

The End of the *Cressy* Years

If Tom's and Angela's 1947 voyage along the Welsh Canal ended ingloriously among the Ellesmere weeds, it did nothing to quench Tom's zeal to pilot *Cressy* across the Pontcysyllte Aqueduct. By April 1949, it appeared that a flow meter was at last in operation at Llantysilio to regulate the amount of water entering the canal from the River Dee, which obviated any need to bother the LMS with requests for blocking off the by-pass channel at Hurleston. Cécile Dorward and Christopher Grundy, moreover, had taken their respective boats *Phosphorus* and *Heron* along Tom's proposed route already, 'the former to Pont Cysylltau [sic] and the latter to Llangollen'.[1] To hear some weeks later that *Phosphorus* had become immovably wedged in one of the Grindley Brook locks was annoying since it delayed his departure and, writing to Robert, Tom was outspoken both about Mrs Dorward's foolhardiness in taking 'so large a boat up the canal when she knew the state of the locks' and her inexcusable abandonment of *Phosphorus*, blocking the canal to all traffic.[2]

Undaunted nevertheless, on 21 May 1949 he and Angela set off from Banbury for what would be a lively trip. Their Banbury-based friends, Bill Trinder and Jim Russell, joined them for a short cruise on the Stratford-upon-Avon Canal between Hatton and Waring Green. Later, skirting Wolverhampton on the Staffordshire and Worcestershire Canal, the Rolts watched a display by the Wolverhampton Aero Club from *Cressy*'s deck. At Wheaton Aston, on the Shropshire Union Canal system, Tom got into a long conversation with the local postmistress which led to her showing him over her own boat. When the Rolts reached Gnosall, Polly Elwyn Jones – wife of IWA supporter Frederick Elwyn Jones MP – arrived by train to have lunch on board *Cressy*.[3] Between Audlem, where it rained hard, and Swanley on the Shropshire Union Canal's Welsh section, which Tom knew as the 'Welsh Canal', they travelled along with a Mr and Mrs Whitley in their boat *Prince*. At the entrance to Baddiley Top Lock, *Cressy* stuck just as she had the previous year and Tom repeated the expedient of dropping the water level to free the boat. On 5 June, having reached Wrenbury Hall Bridge, they encountered Cécile and Alan Dorward aboard *Phosphorus*, now safely clear of the lock. About a week later, since Tom had to go London, he took the opportunity to collect his car from Banbury.

On his return, he brought with him Welsh actor Hugh Griffith who spent a night on board *Cressy*, now moored at Bettisfield. Griffith had befriended Tom after reading *High Horse Riderless* which, unlike the Willans', he greatly admired and in the course of their friendship, he persuaded Tom to join Plaid Cymru.[4] Planning to travel on to Anglesey, when Griffith learned about the plan to cross Pontcysyllte he was keen to take part; soon, he would return to *Cressy* with his wife Gunde.[5] The Rolts, meanwhile, travelled on to Hindford where they sought out Mr Beech, *Cressy*'s builder, who came on board with his wife 'beaming with pleasure' to be re-acquainted with the boat that he had constructed years previously.[6] They reached Froncysyllte, which Tom customarily spells 'Vroncysyllte' on 15 June, where they moored 'outside Mr Davies's cottage' around teatime.[7] Tom, incidentally, took much encouragement from the canal's improved condition which he attributed explicitly to IWA campaigning.[8]

On 26 June, triumphant in the *Cressy* log Tom wrote, 'On this day I realised ambition of taking *Cressy* over Telford's great aqueduct Pont Cysyllte'. From the towpath high above the River Dee, Angela took photographs of *Cressy*, a jubilant Tom, Hugh and Gunde who had recently joined them and were now aboard with Gunde's corgi, ears pricked, on the foredeck.[9] Having achieved their objective, they turned the boat and stern hauled her a short distance up the Llangollen Arm where they found a shady mooring. They spent the rest of the day 'bathing in the river and canal'.[10]

Although it was one of the more companionable episodes of the summer for Tom and Angela, tellingly Tom suggests that it was around this time that their marriage began to fracture. Of the reasons for its disintegration, beyond reference to Angela's parents' continued hostility and her wish to travel abroad while he preferred to remain in Britain he says nothing, except to observe enigmatically that there 'are some things which a biographer, writing in the future, may nose out if he [sic] can'.[11] Nosy biographers can easily make mistakes, but it may be worth observing that as the summer of 1949 advanced, the Rolts often went their separate ways. Admittedly at the beginning of July they were together in the coastal village of Abersoch but, later, Angela went to Venice while Tom remained on board *Cressy*. Often alone, he used the time to revise *The Inland Waterways of England* in the light of Charles Hadfield's observations on the draft.[12] But Trinder, Russell and the engineer David Curwen all visited him at different times, and he took each of them across Wales in the Alvis to see the Talyllyn Railway. His interest in the line remained as keen as it had been on the damp day in 1943 when Angela and he had arrived at Towyh hoping to catch a train and ended up walking to Abergynolwyn.[13]

When the Bill which would become the 1947 Transport Act had been going through Parliament, curiosity had led Tom to examine the list of independent railway companies that were about to be nationalised. The fact that it made no mention of the Talyllyn surprised him. The omission had lodged in his memory and he came to cherish the idea that this one independent railway might survive into the era of nationalisation. The dilapidated locomotives and neglected track hardly encouraged such speculation, but it occurred to him that it might lend itself to reconfiguration on a small scale, not unlike the Ravenglass and Eskdale Railway which he had visited some years previously. Conceived as a mineral line, after the closure of the quarries that it had been built to serve, it had taken on a new lease of life as a miniature railway, running with scaled-down locomotives on a track of 15 inch gauge. To Tom, the idea that similarly small-scale locomotives might run on the Talyllyn line, while the original engines – *Talyllyn* and *Dolgoch* – were preserved in the narrow gauge museum at least merited investigation and with Trinder's and Russell's encouragement, he had approached model railway supplies company Bassett-Lowke for advice. They referred him to David Curwen who lived at Baydon in Wilshire, by chance not far from Aldbourne, and when Tom speculatively visited him, the two men took to one another immediately. Off the cuff, Curwen had estimated that modifying the Talyllyn line would cost 'around £10,000'.[14]

Trinder arrived at *Cressy*'s mooring on 7 August 1949, and Tom drove him to Towyn's Wharf Station the following day.[15] It was probably the occasion when they met the feckless, reckless young man whom Tom mentions in *Railway Adventure* and who had somehow been given the task of driving *Dolgoch*. He caused them some dismay by 'improving substantially on the scheduled running time' between Towyn and Abergynolwyn and, yet more alarming, when they had reached Abergynolwyn station, instead of pausing at the water column to fill *Dolgoch*'s tanks, he drove up the mineral extension to the foot of the quarry incline, skirting a rocky shelf some considerable height above the Nant Gwernol ravine along the way. In answer to Tom's and Trinder's anxious questioning about why he had put himself and his locomotive at such risk, he insouciantly explained that he had wanted to see 'whether the rails were still there', which led them to reflect that he might have satisfied his curiosity by safer means. Unsurprisingly, it was not long after this incident that he left the railway company's employment, having inadvertently broken *Dolgoch*'s frame at the height of the tourist season.[16]

Unaware of this disaster, early in the following month, Tom took Curwen to see the railway.[17] Discovering that, with *Dolgoch* out of action, there could be no question of their travelling by train, the two men explored the recently closed Bryn Eglwys quarries. Not long afterwards, Tom would chance upon a nostalgic article about the Talyllyn line in the *Birmingham Post*. Reading and reflecting on it led him to abandon the miniature railway plan, on the basis that it would 'wholly destroy' the line's character. Instead, making a new, ambitious plan, he impulsively wrote to the paper announcing that there was a 'scheme afoot to ensure the future of the railway'.[18]

As summer gave way to early autumn, Angela returned from Italy, and on 21 September 1949 she and Tom set off from Froncysyllte and headed for Ellesmere. Travelling with them was one Geoffrey Tower who may have been one of the Orreds' Cheshire connections. He evidently enjoyed himself because on 3 October he returned to join the Rolts with his wife Maria for a day's cruise from their Ellesmere home to Fenn's Moss and Fenn's Bank.[19] Only a fortnight later, much to Angela's shock, news came of his sudden death.[20] Since Tom was unwell, she went alone to Tower's funeral which apparently 'was made more sombre by pouring rain.' No sooner had she returned to *Cressy* than she announced that she would be returning shortly 'to Ellesmere [...] at the request of the bereaved to administer further solace'.[21] From the tone in which Tom relates this news – he was writing to Aickman at the time – it sounds as though he did not entirely see the need for this additional mercy dash. As usual, he was busy, preparing for 'the Huntingdon weekend;' he had a meeting in Banbury 'to discuss the railway project;' and he had also to collect his car 'in time for our Saturday departure'. In passing, he complained that he had a 'queasy' stomach and had not eaten a square meal since last he was last with Robert and Ray.[22] The letter gives the impression that he felt under pressure from a number of conflicting commitments, both Angela's and his own. The Aickmans had recently invited Angela and himself to a party which, Tom insisted, they could not attend: 'a trip to London' he wrote 'is really out of the question'.[23]

But Tom always made time for his own projects. During the summer at Froncysyllte, it had occurred to him that the IWA might organise what he described as a 'Rally of Boats.' Market Harborough suggested itself as a location for the event, being accessible both to narrow boats – by branch canal from the bottom of Foxton Locks on the Grand Union Canal – and to broad beam vessels from the Trent. Early in November 1949, Tom and Angela and Robert and Ray made a joint visit to inspect the basin and meet members of the Market Harborough town council. It was, Tom recalls ominously, 'the last occasion on which the first Chairman and first Honorary Secretary of the Inland Waterways Association cooperated together'.[24]

At the time, he was quick not only to draft regulations, a list of awards and an entry form for the event, but also to raise some practical points. He proposed, for instance, that the Docks and Inland Waterways Executive be requested to issue special through passes to enable boaters to attend without having to pay exorbitant toll charges; reminded Robert of the need to determine a fair entry fee which would depend 'on the toll question' and, while he was not in favour of opening the Rally to canoeists, invited the IWA committee to consider whether they should be eligible to take part.[25] On 1 December 1949, he wrote again to Robert asking him to confirm arrangements for a further visit to 'Harborough basin' on the following Monday; returning a newspaper article about the 'grim' state of the Upper Avon Navigation and, in passing, mentioning that he and Sonia Smith, provided that she was available, hoped to call on Professor Schonell of Birmingham University the following Saturday to discuss what he terms 'the boaters' education question'.[26] If his letter was brief, it was nothing less than affable. Yet in the same week, he had evidently tendered his resignation from the post of Honorary Secretary to the IWA; a decision which Robert received with 'consternation and despondency'.[27]

Seeking to explain himself, Tom replied that all along, it had been his intention 'to carry on until the Association was firmly established and then drop out'. 'I think', he continued:

> 'you find IWA activities more congenial than I do because on your own admission [...] you like being what is popularly called a "public figure." This is perfectly understandable and any satisfaction you get in this way is certainly well earned but you see it is a position which I find most uncongenial. [...] What I need to counterbalance my literary activities is some practical and manual activity. The work of the IWA is not a good complement to literary work in my case; in fact, it is quite the opposite.'[28]

The statement gives every sense of Tom's trying to be tactful, seeking to mask his increasing distaste for Robert's style of organisation behind references to his own aversion to finding himself in the public eye. He also shows a polite if somewhat strained reticence over mentioning money, which was in fact, the nub of his trouble. Specifically, he had realised that he could not afford to continue in his post. If he were to make a success of his 'design for living', he had to have time in which to write, since – apart from Angela's small and unpredictable allowance – it was the couple's sole source of income. In stark terms, his unpaid work for the IWA took more time, and, indeed, cash than he could spare. The campaigning trips that he had made would never have been compatible with a full-time job, but as 'a free-lance' he had come to recognise that he was 'always under pressure from others who assume that his time is his own'. He had evidently reached his decision in the light of discussion with Angela. She had been quicker then he to recognise that while they remained active in the IWA, they had little freedom of action, '*Cressy* and her crew' having 'become, by almost imperceptible degrees, the tool of the Gower street office'.[29] Only recently, while Tom had spent 'an enjoyable if hectic weekend' in London with the Aickmans, Angela had declined to join him on the ground that she was 'having an economy drive'.[30] Husband and wife appear tacitly to have decided that stepping back from IWA activity was crucial to the survival of their marriage.

Nevertheless, Tom agreed to remain in his post until the end of the year. Besides planning the Rally of Boats, the Association had to deal with the trouble that Willans had caused. It put Tom in a dilemma. While he had no illusions about Willans' contrariness and was uncomfortable over his courtesy uncle's behaviour in the IWA, he had genuine respect for Willans' skills, appreciated Willans' help in securing his early engineering experience and apprenticeship and felt lasting gratitude for his generosity over the sale of *Cressy*. Meanwhile Robert – exasperated with Willans' insulting letters and increasing tendency to stir trouble within the IWA – had consulted IWA solicitor Sidney Davis about finding a means by which the organisation could expel him. Robert recognised that if the IWA were to oust Willans, Tom and he would have to show a united front. Therefore, he insisted that Tom should attend a meeting scheduled for discussion of the matter, to be held on 11 February 1950 at Newbury.

For Tom, it promised not only to be disagreeable but was also inconvenient. Getting to Newbury from *Cressy*'s Gayton mooring would involve making a long, tedious train journey from Blisworth. Besides, it was his fortieth birthday, and Angela and he had arranged to meet Mima in Stratford-upon-Avon to go the theatre. Adamant in the face of his objections, Robert nevertheless insisted on Tom's attendance.

Frustrated, Tom fired off a furious letter. 'If I do not attend the Newbury meeting', he wrote:

> 'You will make it clear to the others that I am afraid to face Willans and that you are facing the music unaided while I enjoy myself. As you must know [...] I am not in the least concerned by the prospect of meeting Willans [...]. Moreover, I have not the least concern for my personal reputation whatever the outcome may be. What I am not going to do is to go to Newbury as your stooge to act as your yes-man and bodyguard.'

By way of conclusion, he indicated in blunt terms that it would be best for Robert not to come to Newbury, but to leave him 'to handle the situation' himself. 'You need have no fear', he concluded 'that I shall give way to Willans' demands.'[31] But Robert preferred to speak to Willans himself, and insisted that his immediate removal was crucial to the wellbeing of the IWA.[32] The trouble would eventually resolve itself with the discovery that Willans' IWA membership had lapsed.[33]

Ray, meanwhile, had not forgotten Tom's earlier suggestion that Robert enjoyed being in the public eye. Making out that it was nonsense, she suggested that Tom's thinking 'on all subjects' was clouded on account of his and Angela's being 'patently unhappy – a state of mind', she asserted, 'which makes the rest of life unbearable' and which, she continued, led them both to think that 'everyone's hand [was] against them'.[34] 'It is not for you', wrote Tom in reply, 'to pass judgement on our personal affairs and thereby to invent a situation which you believe has warped my judgment. Such personal situations, even if they arose, would not blind me to what I consider to be right and what I consider to be wrong.'[35]

While it is a measured response, he makes it quite clear that her interpretation of events was both otiose and misconceived.

As a canal campaigner, he was acutely aware of the way in which the work practices over the years had given the canal system its specific character. This 'subtle patina of constant use' showed itself in the scars and marks of wear upon towpaths, lock gates, bollards and bridges. Many IWA members, while they might 'value canals aesthetically as an important contribution to landscape', failed to recognise that their distinctive and beguiling character arose from their practical 'utility'.[36] It convinced him both that the IWA should focus its attention on primarily preserving canals which remained in commercial use and that the Association had a special obligation to look after the interests of working boatmen and their families.

Not only did he meet the educationalist Professor Fred Joyce Schonell in December 1949 to discuss the school provision available to boaters' children, but the previous September he had also had a meeting with Joseph Blewitt JP, Midlands Area Secretary of the Transport and General Workers Union, to consider formation of an 'Inland Waterways Branch' of the T&GWU to look after the boaters' interests.[37] Sonia Smith joined him on this visit, her first-hand experience of living on the canals, not to mention her left-leaning politics, making her a useful ally and companion. While Tom did not usually have much time for trade unions, Blewitt seems to have impressed him. Telling Robert about the meeting, not only did he say that it had been 'worthwhile' but alluding to Blewitt's role within the T&GWU, added that 'it was encouraging to find someone in that position with the right ideas.'[38] If it was this meeting that he wrote about so scathingly in *Landscape with Figures* – describing it as a 'frustrating interview' and complaining that the union official 'just did not want to know about [boaters'] problems' – it is difficult to account for the shift in perspective. Nevertheless, it is entirely believable that a shared mission designed to interest the trade unions in the needs of the working boatmen should have drawn Sonia and Tom together. How literally we should take Tom's account of their mutual recognition, while they drank tea together in a 'squalid Birmingham café', that they were in love with one another, is open to speculation, although the drab setting certainly serves to heighten the drama of the revelation.[39] At the same time it feels likely that their attachment to one another may have grown slowly over the months – years – in which they had come to know one another. It also seems possible, probable, even, that their shared feelings for one another became apparent to other people before either of them realised what was happening.

Plans for the Market Harborough Rally of Boats developed stormily. Disagreement broke out over the trophies donated for the event and, despite palpable reluctance on her part, Angela found herself inveigled into arranging for a fair to be included in the festivities. Most provocative in the Rolts' view was Robert's insistence upon staging Benn Levy's comedy *Springtime for Henry* during the event with either *Allowance for Error* by Robert Vigo or Alfred Sutro's *A Marriage Has Been Arranged* as a curtain raiser.[40] In *Landscape with Canals*, Tom maintains that his aversion to staging the plays at the Rally of Boats stemmed from their having 'no [...] relevance so far as the waterways were concerned' and serving only to distract the committee from more important matters.[41] In fact, they invited objection on other grounds. Levy's piece, which Artemis Cooper tellingly describes as a 'sexy little farce', turns upon adultery.[42] Since Robert, married to the devoted Ray, was himself in a relationship with Jane whose marriage to Peter Scott had recently ended and while Jane herself had observed that Tom was 'very much in love' with Sonia, it must have bitten rather close to the bone.[43]

In May 1950, mindful of Harry Batsford's planned 'book about the Thames', Tom and Angela took *Cressy* up the river from Teddington to Lechlade. Although Tom had seized on Batsford's commission and was most appreciative of the Thames Conservators' generosity in waiving their usual licence fee and lock dues, writing *The Thames From Mouth to Source* does not appear to have brought him much satisfaction. Staines he found a 'dreary town'; Marlow a 'snobs' corner' and the Eton College oarsmen caused a 'frightful delay' at Boveney Lock. 'Why don't they use rollers?' he demands in the *Cressy* log.[44] Lacking the raw anger of *Narrow Boat* and *High Horse Riderless*, *The Thames From Mouth to Source* rings with Tom's various grievances. Modern boats, he asserts, are no more than 'aquatic motor-cars'; the 'suburbs of swollen Oxford' have enveloped Osney; there is an 'appalling collection' of 'caravans and shacks' at Hurley; while Bablock Hythe bears such 'unsightly signs of weekend popularity' as 'a row of "chalets"' and an old bus, presumably used to provide living accommodation.[45] In *Landscape with Canals*, when Tom recalls drafting his book about the Thames, he suggests that some readers may discern 'a certain sad, elegiac quality about the writing' which he attributes to his 'secret' awareness that it was to be the 'last summer' of his and Angela's 'design for living.' Perhaps he is too generous to his younger self. Arguably readers are more likely to find the book irritable and querulous, if not downright snobbish.[46]

Once Tom and Angela had completed their survey of the river, they found a sequestered mooring for *Cressy* near Lechlade in easy reach of the town. Whatever tension was growing between Angela and himself, Tom wanted nothing more than to spend the summer there, much as he had spent the previous summer at Froncysyllte. When he was not working on his book, he explored Kelmscott Manor and forged a friendship with John Betjeman who visited nearby Buscot to open a fete.[47] As for the politicking of the Council of the Inland Waterways Association, he had had more than his fill of it. Thinking that he might give the Market Harborough rally a miss, in June 1950 he definitively resigned both from the IWA Council and as secretary. Writing to Robert from Lechlade, he not only insisted that no publicity be given to his withdrawal from IWA activity but ended his letter with the statement that he had informed Angela that Robert had recently made an 'irrelevant and unforgiveable reference' to her.[48] Precisely what had happened is not clear but soon afterwards, Ray asked Angela for repayment of a loan. Angela was surprised, since at the time, she and Ray had agreed that there was no urgency in the matter, and that she understood that Ray would not expect repayment until she – Angela – came into her inheritance.[49] In the event, Tom settled the outstanding sum. 'Words fail me' he remarked.[50]

Nevertheless, when Robert specifically asked him not to come to Market Harborough, it was a surprise. By chance, the week of the Rally of Boats – 14–19 August – was precisely when Tom's *Inland Waterways of England* was due to be published. Although the opportunity to use the event for publicity may not have occurred to Tom previously, the chance of timing had not escaped his publishers Allen & Unwin. When Philip Unwin, with whom Tom was on excellent terms, gleaned that the IWA opposed any plans for publicising the book during the rally, he wrote to Tom in some puzzlement to ask what was going on. Startled, Tom changed his tune.[51] Nothing, he resolved, would stop him from going to Market Harborough. 'When [...] you say that nothing but harm to the Association can result from my appearance at the Rally', he informed Robert from *Cressy*'s Lechlade mooring:

'I am forced to assume either that you imagine that I am going with the intent to sabotage the success of the event [...] or the remark carries a veiled threat that me coming will 'force' you to create an unpleasant situation.'[52]

It was fighting talk. Guessing that Robert might have taken some mischievous pre-emptive measure to stymie his sales, Rolt and Unwin between them may have employed some clandestine measures to encourage various Midlands newspapers to publicise the Rally and the book in the same breath, or at any rate, in the same columns. Between 1 and 11 August 1950 the *Leicester Evening Mail*, the *Banbury Advertiser*, the *Market Harborough Advertiser and Midland Mail* and the *Northampton Mercury* all gave extensive coverage to the imminent appearance of *The Inland Waterways of England*, while also drawing 'by-the-way' attention to the Market Harborough event.

The coming-together of boats from all over the country proved immensely popular and between 14 and 19 August 1950, more than 50,000 visitors flocked to see them.⁵³ It was also a chance to watch the plays; enjoy Ealing Studios' 1945 film *Painted Boats* for which Tom had acted as a technical adviser, which was shown together with the British Transport Commission's *Inland Waterways* in the local cinema, attend an exhibition of Peter Scott's paintings and take trips on George and Sonia Smith's working boats *Cairo* and *Warwick*. Stalls sold canal-related art and artefacts – everything from decorated cans and dippers to Brindley and Telford headscarves. There was a Trade exhibition; a lecture by Robert; a cricket match; music from a silver band; election of a festival queen and a resplendent fireworks display laid on by Market Harborough Town Council. On the closing night, there was a grand ball in the Assembly Rooms and presentation of trophies.

Despite the resolution with which he had defied Robert's request not to attend, Tom may have been more anxious than he cared to admit. For much of their journey from Oxfordshire to Leicestershire, the Rolts had had the company of the Humphries family but when they reached their destination, they found that they had to moor some distance from the heart of the proceedings, while Peter Scott's *Beatrice* took pride of place, serving as the rally's waterborne headquarters. In the published version of *Landscape with Canals*, Tom admits to being aware of 'hostile eyes peering' at his boat from behind *Beatrice*'s window curtains leading him to surmise that 'had *Beatrice* been equipped with torpedo tubes, *Cressy* would have gone straight to the bottom'.⁵⁴

Although he was wary, Angela made every effort to preserve cordial relations both with the rally organisers and the townspeople. If Robert is to be believed, she made a number of visits to the rally organisers' office in Welland House to 'say something kind' to them.⁵⁵ She agreed to pose 'doing boat chores' for a photograph which appeared in the *Leicester Daily Mercury* which, probably on account of Mr Fortune's editorial influence, always took an interest in the Rolts' life aboard *Cressy*. In the picture, she is swabbing *Cressy*'s deck- wearing a striped skirt made 'from pillow ticking' – a tightly woven linen or cotton which apparently was 'easily washed [and] ironed and not too flimsy for the waterside'.⁵⁶ Unable to resist the lustre of a famous name, the journalist who wrote the piece added that at the time when the photograph was taken, the Rolts were about to entertain the poet and critic Sacheverell Sitwell; years after the event Robert, who evidently had an infinite capacity for nurturing grudges, complained in print that Tom did not introduce him.⁵⁷ After the Rally had finished Angela wrote to the local paper to express her gratitude to the people of Market Harborough, 'most especially', she said, 'to the kind person who did […] my washing and ironing for me and the two boy scouts who [….] helped with any job about my boat that they were asked to do'.⁵⁸

Even if Tom grew increasing melancholy, he drew some comfort from a visit from A.P. Herbert – Sir Alan Herbert, the IWA President – who braved Robert's indignation to come aboard *Cressy* at Market Harborough for some 'drinking and yarning'. The coincidence of Herbert's having highlighted the stigma of marital breakdown and the cumbersome requirements of the divorce laws in his 1934 novel *Holy Deadlock* could hardly have escaped Tom's notice. Whether the novel or its subject-matter came up in the course of their conversation is unknown, but it is believable that at the time Tom found Herbert an ideal companion and confidant.[59]

From Market Harborough, the Rolts went to the Norfolk Broads to spend a week with Brigadier Hopthrow and his family on their wherry, *Dragon*.[60] Hopthrow was not only an IWA member, but also held a senior post with ICI. Around the time of the Lifford Bridge campaign, he, Tom and 'Mr Mallender [...] of the Derby Gas, Light and Coke Company' had attempted to bring a couple of coal boats along a neglected part of the Derby Canal. Somehow, news of the plan had reached the authorities who promptly foiled it by padlocking lock gates along the way, but Tom and Angela must have remained on friendly terms with Hopthrow's family since *Dragon*'s log mentions their staying on board the wherry over 22–28 August 1950.[61] *Dragon* – built around 1900 – appears, rather like *Cressy*, to have started out as a trading vessel before conversion for leisure usage. Not only did the Hopthrow family spend much time on the boat, but they also took part in wherry races. It may well be symptomatic of Tom's unease at the time that he passes over his and Angela's visit to the Broads in silence which begs some question of whether it marked further breakdown of the Rolts' marriage. At the end of their holiday, the Rolts returned by stages to Banbury, their accustomed winter mooring, for what Tom describes as 'the most unhappy winter and early spring' of his life.[62] In a small, but perhaps telling detail, from November 1950, he adopted Trinder's address – 84 High Street, Banbury – in place of MB Cressy, Tooley's Boatyard, Factory Street, Banbury.[63] He does not disclose whether he was lodging with Trinder at this time or treated his premises only as a 'poste restante', but two external factors, he explains, combined to put his and Angela's marriage under increasing pressure.

First, in November 1950, wishing to present a clear statement of policy, the IWA Council defined the Association's *de facto* mission as being 'to advocate the restoration to good order, and maintenance in good order, of every navigable waterway by both commercial and pleasure traffic'.[64] In public-relations terms, the objective made a ready appeal. Anyone with a fond regard for the nation's waterways might welcome the notion that they should all be open to all forms of traffic and fit for purpose. What was more, the policy as stated promised to draw any and every new leisure-minded boat owner – many of whom, like Peter Scott, were wealthy and influential – into the IWA fold. While some members might share Tom's wish to see the IWA direct its efforts to maintaining the portion of the canal network that remained in commercial use, that option may not have held the same broad appeal with the public at large as Robert's wish to 'save every single mile'.

Having by this time distanced himself from the IWA, Tom might have let the matter pass him by were it not that Rendel Wyatt, the pioneer of canal hire boats from whom he and Robert had chartered *Ailsa Craig*, called on him one winter's night at Banbury to draw his attention to another Aickman initiative, namely a significant alteration to the IWA's constitution. Robert, it appeared, had instigated a change of rules; specifically it meant that any member who did not agree with his new policy would be requested to resign his or her membership.[65]

By now, Tom had little stomach for a fight. The prospect of allowing himself to be seen as the 'spearhead of a dissident movement' within the IWA caused him much misgiving and he had no wish to compete with Robert for influence with the members. In truth, he wanted only 'to have done with the whole silly business'.[66] What swung his decision to join Wyatt in contesting Robert's plans was his awareness that the IWA included a number of people whom he regarded as close friends and who would not have joined the Association, had he not been active within it. Cavilling at the pettiness of Robert's rule revision and concerned at his proposed change of IWA policy, Tom agreed to circulate a memorandum to all members. It drew attention first to the 'dangers' inherent in Robert's proposed alteration to the Association's constitution, and secondly, it 'stressed that the IWA must [...] do more to help the working boatmen'; notably, give more attention 'to that part of the narrow canal system which was still in commercial use'.

A meeting of IWA members held in Birmingham on 30 December 1950 agreed by 233 votes to 26 that all waterways should receive IWA support, rather than only those that carried commercial traffic.[67] Nevertheless, when the proposed rule-changes came up for discussion, the meeting turned so stormy that adjournment became inevitable. It was agreed to reconvene on 3 February 1951 when, at another tempestuous meeting, Robert's revisions of the IWA constitution was accepted by a majority of 92 votes to 25.[68] For Wyatt, Tom and their supporters, it represented a total rout. At a meeting of the IWA Council a fortnight later, a number of council members including Sonia Smith, Angela Rolt and Rendel Wyatt were conspicuous by their absence. Teddy Edwards, who had succeeded Tom as the Association's secretary, sent out letters on Robert's behalf to request either their resignation or 'recantation'. Years after the event, when Tom was dead, Sonia memorably described this requirement as 'a kind of unnecessary Elizabethan oath of the most hideous and harmful kind'.[69] If the language seems melodramatic, it was of a piece with the Council's next step which was to create an expulsion sub-committee. Under the auspices of this body, signatories of Wyatt's and Tom's memorandum were given two weeks either to tender 'express assurances' of their loyalty, or to resign. In the event, none of them offered any response.[70]

If Angela retained any commitment to the IWA in the months that followed the Market Harborough Rally, it ebbed fast. According to a Mr A.S. Cavender, sometime chairman of the Fenlands branch of the IWA, 'she was in the habit of stating her intention of joining a circus'.[71] Sometime in the spring of 1951 she put this plan into effect, purchased a Morris Oxford two-seater car for £10 which Tom serviced for her, and headed off to take up employment with Billy Smart's Circus, whose headquarters were at George's Stables in Winkfield near her family's home in Sunninghill. Since she was used to horses and had a fair command of French and Italian, she must have been a useful recruit. Years later, she told Ian Mackersey that she 'had a smelly uniform and did all sorts of things like selling programmes and looking after the horses and interpreting because [she] spoke French'.[72] Even if he were in love with Sonia, Angela's departure left Tom numb.

For the breakdown of their marriage, he lays much blame upon her parents. 'I believe', he writes:

> 'that the extraordinary, implacable hostility of Angela's parents towards her marriage, by cutting themselves off from her for good, placed an intolerable psychological burden upon her which I did not fully appreciate at the time....'[73]

Yet his remark about the Orreds' distancing themselves from their daughter does not entirely stand up. Besides mentioning that in the weeks before the couple's wedding, Mrs Orred had clandestinely helped Angela to choose curtain fabrics for *Cressy*, in a letter to Robert dating from 1946, he slips in the detail that Angela had 'arranged to have tea with her mother on Thursday as this is the only time the latter can manage'.[74] If it was another secret assignation, it raises large unanswered questions about how often mother and daughter managed to meet, what they talked about, and, indeed, what Tom had made of these encounters.[75] Angela had some private means – possibly a bequest – to which Tom refers as her 'allowance' and which gave the couple a financial cushion. But they clearly went in dread of Major Orred's negating the arrangement, which invites some speculation as to how it had been set up in the first place. He was certainly not above attempting to rifle her resources. In December 1946, about to move back onto *Cressy* after their spell in the caravan during the boat's repair, Tom told Robert that Angela's father had written to her, 'proposing that he relieves her of £5,000'. Appalled, he – Tom - had advised her 'under no circumstances to commit herself to anything' but to consult solicitors. 'Her father's behaviour', he remarked, had 'always been quite scandalous, unscrupulous and unprincipled'.[76] Even if Major Orred's threats were to prove groundless – which was not by any means certain – they were alarming and left Tom acutely conscious of the need for time in which to write, his publications being the couple's most dependable source of income. Even if the Rolts were in fact better off than many families of the time, nagging money worries cannot have helped to sustain their marriage.

There is also some question about how far the Rolts' childless state was a matter of choice or inevitability. From Tom's books and, to some extent, his letters, Angela and he emerge as a pair of committed adventurers, complicit in their dedication to the canal cause and their enjoyment of theatre and music, with little inclination for family life. While the matter of starting a family is a subject that Tom never broaches, no evidence survives to suggest that either he or Angela shared the Aickmans' outspoken dislike of children, to whom they disparagingly referred as 'wombats', and which they viewed as an 'obstacle to civilised life'.[77] Tom enjoyed the company of the Grundy sons while Angela adored it when, years after she and Tom had gone their separate ways, Tom's and Sonia's sons Richard and Tim visited her in France. To one of her numerous chatty letters to Sonia of the 1970s, she appends the touching message, 'I am glad the Boys are in good fettle'.[78] Whether she and Tom had experienced fertility problems, or shrank from the sheer difficulty of raising a family in the close confines of a boat or chose to devote themselves entirely to their respective careers of writer and canal photographer is a mystery.

The fact that Angela left Tom just when he recognised the full extent to which rot had taken hold within *Cressy* deepened his shock. As far as Angela remembered, the old boat had 'always been in a terrible state' and ever since the day in 1939 when he purchased her from Willans, Tom had had 'to patch up and patch up to keep her together'.[79] Inevitably, there were limits to what this rather desperate-sounding programme of running repairs could achieve. The crew of a working boat would usually reckon to lift the floorings at the end of every journey, so as to clean out the hold. Since the hold of a houseboat did not require cleansing so often, it was easy to make the assumption that the flooring would and safely could remain *in situ* pretty well all the time. Certainly there was no provision for the lifting of *Cressy*'s flooring and, more seriously, there was no means of achieving through-ventilation beneath it. That the aperture between the flooring and bottom of the boat would be just the type of place where dry and wet rot were likely to take hold apparently did not occur to anyone involved in adapting the boat for residential use. With his practical engineering instincts, it seems surprising that Tom did not recognise the potential hazard sooner, but when he planned his 'design for living' the technique of converting working craft for domestic life had been in its infancy. Having once taken stock of the situation, Tom recognised that it was beyond remedy. 'To restore *Cressy* to perfect order', he reflected, 'would mean virtually a new boat in which nothing would be left of the original but the iron knees'.[80] To proceed with so drastic a rebuild, even if he had wished to undertake it, was beyond his means.

By this time, he had no inclination to prolong his life on the canals. Instead, he prepared himself to adjust to what the boaters would call 'life on the bank'. It was, in a manner of speaking, a clean break. Rendel Wyatt, by chance, knew of a prospective buyer for *Cressy* and, since all that this lady required was a houseboat for herself and her father, Tom did not dismiss the prospect of the sale out of hand. Instead, with Mel – Helen Melville Russell-Cooke, daughter of Captain Smith of the *Titanic* and sometime lover of Tom's painter cousin David Rolt – as a travelling companion, he worked the boat over the 105 miles from Banbury to Stone.[81] With so many memories inhering themselves in the landscape through which they travelled, it was a journey of infinite sadness; in the *Cressy* log, he does not even record its date, but writes 'R.I.P' beside the record of the mileage.[82]

Even the presence of the prospective purchaser – her name was Miss Matthews – 'literally waving her chequebook' on the wharf at Stone when they arrived did nothing to raise Tom's spirits. Wyatt who, after all, had a reputation to maintain, insisted that a survey of the boat's hull be made before any money changed hands, and at that point, Tom knew that *Cressy*'s fate was sealed. Sure enough, seeing the rot, the boat surveyor condemned the vessel out of hand. Tom matter-of-factly retrieved his essential belongings and consigned *Cressy* with her fixtures, fitments and gallant decoration to oblivion, although in 1952 he would retrieve her Model T Ford Engine for installation in a maintenance locomotive on the Talyllyn Railway.[83] In the short term, he went to stay with his mother at Stanley Pontlarge, although he kept some footing in Banbury and continued to treat Bill Trinder's radio and gramophone shop at 84 High Street as his effective postal address until June 1951 when he moved to Towyn.[84] But *Cressy* lived on in his thinking. Years after her timbers had been repurposed and what remained of the boat either sunk or been broken up, Tom would dream that he was aboard her once more.

Chapter Eleven

The Talyllyn Railway

'You may or may not be surprised to learn', wrote Tom to Philip Unwin in June 1951, 'that after twelve years I have forsaken canals and gone over to railways'.[1] At the time, Unwin hoped that Tom would write a book about canal legends but the letter, written on paper headed 'TALYLLYN RAILWAY COMPANY, TOWYN, MERIONETH' seems specifically designed to quash such expectations.[2] If Tom had managed to combine his early interest in the line, sometimes rather uncomfortably, with his canal commitments, by the time that he had parted company with the IWA, with *Cressy* and with Angela, he wanted a new beginning. Even if it were not entirely easy to achieve in practical terms, the divide which he draws in the *Landscape* books between the canal and railway phases of his life could hardly be sharper.

The railway owed its origin to slate – specifically, to the quarry at Bryneglwys a little way south east of Abergynolwyn. When Tom brought David Curwen to the site in 1949, it had been closed for about a year. They found ledgers open in the clerks' office, a lathe gleaming with oil and ready for work in the millwrights' shop and a dart board hanging on the wall of the mess room in the quarrymen's barracks.[3] Local man John Pughe had started to quarry the slate of Bryneglwys in the 1840s and, some twenty years later, its quality caught the attention of the Manchester brothers, William and Thomas Houldsworth McConnel. It was the time of the American Civil War and the Union blockade had starved their mills of cotton. Keen to develop new commercial interests, they formed a company for the purpose of working slate and invested some £15,000 in promoting and building a railway between Abergynolwyn and Towyn. It obtained its enabling Act of Parliament in July 1865 and opened to traffic the following year.

By way of motive power, it acquired two state-of-the-art tank engines from Messrs Fletcher Jennings & Co of Whitehaven. While No.1, *Talyllyn*, was of 'orthodox' design, No. 2, *Dolgoch,* boasted an unusually long wheel base in order, the makers promised, to secure 'greater steadiness than can be obtained with engines in which the wheel base is short and the firebox overhangs the driving axle'.[4] Intended to promote stability, this design had the effect of cramping the locomotive's valve gearing – that is, the mechanism which controls the timing and admission of steam to the cylinders – between *Dolgoch*'s narrow frames.⁵ Since it made the mechanism difficult to access, over the years hard-pressed Talyllyn staff were rather apt to neglect its need for maintenance.⁶

The quarries' output never entirely fulfilled the McConnels' expectations. As Tom's friend and fellow Talyllyn man John Snell observed, 'the further the exploratory tunnels were driven, the poorer the slate became'.⁷ By 1900, Bryneglwys was in the ownership of Thomas Houldsworth McConnel's four sons, of whom only one – William Houldsworth McConnel – took any serious interest in its workings. Although, by this time, most of the good quality slate had been removed, Bryneglwys had a spurt of commercial success between 1900 and 1903 when a strike halted production at the vast Penrhyn Slate Quarry near Bethesda. But it was only short-term gain, for the Welsh slate industry was already in decline and by the time that Penrhyn resumed work, the building trade had begun to import slate from overseas.⁸ At the end of 1909, when his leases were about to expire and his quarries were no longer profitable, McConnel closed them down. 'Overnight', Snell relates, Abergynolwyn became 'a village of the unemployed'.⁹ At this point Henry Haydn Jones, Towyn ironmonger and Liberal Party Parliamentary candidate for Merioneth, intervened. Not only did he buy the McConnel brothers' interests in the village property, but he renewed the quarry ground leases which had been about to expire and purchased both the business and the railway for £5250. It helped to boost his standing locally and in the 1910 election, he polled 6085 votes against the 1873 votes for the Conservative candidate, his nearest rival.¹⁰

Although they were a doubtful asset, Haydn Jones – later Sir Henry Haydn Jones, known as Sir Haydn – kept the Bryneglwys quarries open for almost forty years, but a serious rock fall in 1946 forced the mines inspectorate to take action, with the result that 'several [...] chambers were surveyed, condemned and dynamited'.¹¹ A few men remained at the site in the short term to clear the remaining slate that was suitable for sale, but Sir Haydn brought his tenancy to an end, and in 1948 all production at the quarries definitively ceased. The railway, meanwhile, which carried modest local traffic throughout the year and experienced a surge of tourist passengers every summer, remained precariously in business.

How many times Tom visited it, either alone, or with Angela or with his engineering-minded friends, it is not easy to deduce, but the place clearly haunted his thinking. His later-life recollection in *Landscape with Figures* of the visit that Angela and he made in the war years breezily makes light of their misfortunes.¹² Despite the absence of any train, the wet weather and – until the district nurse takes pity on them – the prospect of a long walk back to their lodgings, Tom at any rate enjoys the outing. The visit that he describes in *Railway Adventure* is more sombre. Written closer in time to the events that it describes, it may be more faithful to the spirit in which Tom embarked on the Talyllyn project. In this account, Tom portrays himself as being alone, a solitary explorer of the line. That is not to say that it is unpeopled. Walking along the track he hears someone working with a heavy hammer in the engine sheds at Pendre. 'The victim of this forceful treatment,' he surmises was [locomotive] No. 1, *Talyllyn*'. He also encounters a pair of aged plate-layers only to realise that they speak no English. Nevertheless, he emerges as the hero of the unfolding drama and it was an impression which would, to some extent, shape his thinking about the way in which the railway developed in the early 1950s.

Not only did his interest in the line lead him to bring Trinder, Russell and Curwen to see it, but also, if they were fortunate, to travel over it. On his visits, he made himself known to the railway's long-serving general manager, Edward Thomas – 'Secretary, Accountant, Booking Clerk, Station Master and Guard all rolled into one' – with whom he came to enjoy close friendship.[13] Aware of Bryneglwys's historical links with the railway, he made arrangements for a representative of the Rural Industries Bureau to inspect the quarries to see if there was any chance that they might return to profitable business. As it turned out, the visiting inspector's report was less than encouraging and clearly indicated that the railway could expect no future slate traffic.[14] Then, in September 1949 he chanced upon a piece about the railway in the *Birmingham Post*, its unknown author celebrating the beauties of the line, lamenting that it was 'obviously on its last legs' and expressing that hope that someone – 'the Government or British Railways' – might intervene and come to its aid. It was though fate had played into his hands. Seizing his chance, Tom promptly wrote to the paper himself, emphasising that without the injection of 'practical and financial help' it was 'extremely unlikely' that the railway would open over the following summer. He announced that there was 'a scheme […] afoot' to ensure its future and urged interested members of the public to make contact with him so that he could inform them of future developments. The replies, which reached him in great number, he filed away for future reference.[15]

It was actually Trinder who took the lead in negotiations with Sir Haydn concerning the railway's future. He, after all, shared Sir Haydn's commitment to the Liberal Party in politics and the Methodist Church in religion, which gave the two men some common ground. By November 1949, they had evidently had some exchange about the railway, for Sir Haydn wrote to Trinder, suggesting that, rather than travel all the way from Banbury to Towyn, he might care to outline his 'scheme' in writing.[15] Trinder replied the following month, after consultation with his 'friend, Mr Rolt.' Rolt, he reminded Sir Haydn, had once accompanied him on a visit to Towyn. It had probably taken place in the autumn of 1949 and it was Tom's only encounter with the railway's owner.[16] Tom remembered the occasion well. Sir Haydn had spoken about the railway 'without sentiment and to the point'. It was, he had said, losing him money, but he would continue its summer service for as long as he lived. While he would welcome any proposal for prolonging its working life, he foresaw that it would require massive investment 'with little prospect of any commensurate return'.[17]

Tom at this time had not yet abandoned the idea of reconceiving the Talyllyn as a miniature railway and Trinder began his reply to Sir Haydn by describing how, small-scale wise, it would work in conjunction with the existing line. More significantly, he was at pains to emphasise that Tom and he were serious in their wish to keep the railway running. They were neither, he explained, 'men of means' nor speculators seeking to make a profit. Their hope was to preserve the Talyllyn Railway for the benefit of enthusiasts; indeed, they foresaw that any financial aid that reached it was far more likely to accrue from visitors' small donations than from anything resembling a capital investment. Furthermore, Tom and he had no wish to see it 'run by outsiders' but hoped instead that 'it should remain [...] intimately connected with the life of Towyn and district'.[18] While Sir Haydn clearly replied – in a subsequent letter, Trinder specifically thanks him for his response – his answer does not appear to survive. A conscientious campaigner and correspondent, Trinder wrote to him again in March 1950 saying that the railway had been 'much in the minds of many people recently' and offering to acquaint him with Tom's and his own current thinking about its future. More specifically, he asked if Sir Haydn would allow them to call a public meeting 'probably in Birmingham' at which he envisaged sharing plans for the railway with a gathering of enthusiasts, with a view to exploring their practicability.[19]

Caught in the short term between organisation of the IWA Rally of Boats and planning his book about the Thames, Tom temporarily put the Talyllyn to the back of his mind. But early in July, when Angela and he were on their way to Market Harborough, he heard that Sir Haydn had died.[20] In haste, he wrote to Sir Haydn's executors and the solicitor charged with administration of his estate, to request that they made no 'irrevocable' decision concerning the Talyllyn Railway until he had had the chance to arrange a public meeting to discuss its future and to consider the possibility of forming a society dedicated to the line's preservation.

Sir Haydn's executors, as it happened, were more than happy to follow the advice of Edward Thomas. Since he was entirely of Tom's mind where the future of the railway was concerned, having listened to him, they unhesitatingly gave their agreement.[21] Before long, Tom had established contact with everyone who had responded to his letter in the *Birmingham Post* and a meeting was fixed for 11 October 1950, to take place in Birmingham's Imperial Hotel.[22]

About seventy rail enthusiasts attended, together with Edward Thomas from Towyn and Sir Haydn's daughter, Mrs Mathias. Trinder, who took the chair, conceded that the railway was popular with holidaymakers and offered great possibilities for the future. At the same time, the 'hard facts,' which Sir Haydn had recognised, were that it was losing money, that the entire track needed renewal and the locomotives required repair. His death made these matters all more pressing, and unless 'external help in some form was forthcoming' the likelihood of the railway's reopening was small.[23] Tom, at this point, indicated the need for a working committee to decide whether the best approach was either to purchase the railway, or to form a society to keep the line running under its existing ownership. Whichever approach the meeting chose to adopt, it had to give immediate priority both to the track, which was in a shocking condition, and to the need for additional locomotives to enable the railway to open for six days a week instead of only three.

Edward Thomas reported that quantities of rail and sleepers previously belonging to the nearby Corris Railway – it had closed after a bridge collapse in 1948 – were available for purchase. His assurance that the railway had made a modest profit the previous summer and had more passengers wishing to travel than it could accommodate placed the whole project in an optimistic light. Indeed, according to Talyllyn supporter and railway photographer Patrick Whitehouse, Mr Thomas's words imbued the entire meeting with a 'spirit of crusade'.[24] Whitehouse was one of the 'hardened and somewhat sceptical' enthusiasts – members of the West Midlands branch of the Stephenson Locomotive Society and/or the Birmingham Locomotive Club – who were known collectively as the 'Birmingham Railway Mafia'. Later, he recalled how, in the mood of '"you can't do it" or "you can't have it" which tended to characterise post-war Britain,' Mr Thomas's confidence that saving the railway was possible and plausible made immediate and immense appeal to everyone who heard him.[25] From the floor, a Mr Grey testified to the Talyllyn's popularity, reporting that on each of his three recent visits, he 'had had to struggle for a seat'.' Other contributors expressed their confidence that the line would be able to recruit volunteer labour, emphasised the importance of publicity in catching and retaining public support and pledged their readiness to serve on the working committee.

A fortnight later, the working committee held its first meeting and voted to adopt the name of the Talyllyn Railway Preservation Society. Bill Trinder found himself elected Chairman, with Whitehouse as Secretary. Patrick Garland, an accountant by profession who made his office available for future meetings, became Treasurer and Tom took on the role of Publicity Officer.[26] Another ten committee members were elected and a further five – including David Curwen who had recently married Tom's cousin Barbara Willans – were co-opted.[27]

The issues concerning the state of the permanent way and the need for additional motive power required immediate attention. The condition of the Talyllyn track was rough beyond words and caused frequent derailments.[28] Following the 1948 cessation of operations on the nearby Corris Railway, the Talyllyn – as Edward Thomas informed prospective supporters at the meeting of 11 October 1950 – had purchased 'ten tons' of Corris rail.[29] The remainder had been snapped up by an eager local scrap merchant. Aware that they had their metaphorical backs to the wall, the Talyllyn party opted to buy from him all that they could afford, although, as Tom ruefully reflected, 'at the price demanded, this could be only a tithe of the quantity needed'.[30] Although they did not have enough Corris rail to complete the work, the resourceful Talyllyn supporters managed to source additional rail and sleepers from such far-flung locations as Nuneaton, Crich and Betchworth in Surrey.[31] In the event, improvements to the permanent way would occupy much of the following twelve years, completed largely with volunteer labour and some assistance from the Territorial Army.

The society's other concern was to find additional motive power to provide back-up for *Dolgoch. Talyllyn*, the railway's other locomotive was, by this time, worn out – a 'forlorn and rusting hulk' – which, from their inspection, the Preservation Society's early members considered well past operable use and best left in a barn, abandoned by all but the hens who laid eggs in the smokebox.[32] Again, Talyllyn eyes turned to the Corris Railway which shared the Talyllyn's unusual 2ft 3in gauge and by now had two redundant locomotives for sale which – most advantageously in terms of economy – would not require re-gauging.

That was not to say that their acquisition went through without careful thought. When Tom went with Jim Russell and David Curwen to Machynlleth where the engines were stored, their rusting state and smashed boiler-gauge glasses caused him momentarily to lose heart. But Curwen's confidence that the appearance of decay was no more than surface deep and, significantly, that the boiler of the older engine of the two appeared to be 'in fair order' offered grounds for optimism. Known informally as 'the *Falcon*', Corris Railway No. 3 was a product of Henry Hughes & Company's Falcon Works in Loughborough and had come into service in 1878. Corris Railway No. 4 was one of Kerr Stuart's 'Tattoo' class, a detail which held powerful significance for Tom. Dating from 1921 she was significantly the younger of the engines, but had evidently come in for much harder work than her companion; Tom, indeed, would wryly observe that she had been 'worked almost to death'. Curwen reckoned that while her boiler might need remedial work, it was not beyond repair and Tom recalled that Tattoo locomotives, being intended for contractors' use on large-scale construction sites, were designed on 'simple lines with a view to ease of maintenance' and straightforward replacement of outworn parts.[33] The advantages of purchasing the Corris engines were too great to dismiss, although the need to pay for crucial track repairs put the £65 asking price for each locomotive beyond the Preservation Society's limited means. Fortunately, negotiation with British Railways at Swindon – stronghold of the GWR, the Corris's current owners – resulted in the Talyllyn's purchase of both engines, if not precisely for a song, then certainly 'for less than the previous price of one'.[34] In mid-February 1951, Tom triumphantly informed Philip Unwin that the two locomotives were due 'to be transferred to Towyn very shortly'.[35] In a tribute to the railway's management of the recent past, they would acquire the names *Sir Haydn* and *Edward Thomas*.

In the same month, at Unwin's instigation, Tom lectured on the 'Romance of Canals' at the Ideal Home Exhibition.[36] It was no more than a brief diversion from his railway commitments; the next month, he agreed to take on Edward Thomas's post of general manager of the railway for the coming summer and joked to Unwin about growing a beard in the style of the 'august gentlemen whose pictures one sees in old volumes of the *Railway Magazine*'.[37] 'Never in the course of my lifelong interest in railways had I ever dreamed that I would […] be called to take charge of a real public railway within the meaning of the Act', he reflected later.[38] Ranking his priorities ahead of the Whitsun opening, he determined that the railway should offer a five-days-a-week service to the travelling public from 4 June until September.

Managing the railway staff could be a challenge. With most of them – Birmingham-born Bill Oliver who had married a Towyn lady, for instance, and Bill Maguire known as 'Maggie' who had served with the Indian police force before independence – he got on well. But things were not so easy with father and son Hugh and Dai Jones, two staff who had been with the railways in the Haydn Jones era. Notwithstanding Edward Thomas's observation that they were 'very difficult people', when first he met them, Tom had taken to them warmly, but as the weeks passed, it appeared that the Joneses did not entirely care for the new regime and expressed their dissatisfaction by walking out at short notice, Hugh just before the railway's Whitsun opening and Dai soon after.[39] The reasons for their short notice departure eventually came to light. Hugh, it appeared, was used to driving trains and did not care for Tom's asking him to lead the 'newly-established Track Gang'. That he was the only person around who had any substantial experience of permanent way repair and therefore the obvious candidate for the task counted for nothing with him and he showed his dissatisfaction by walking out. His son, who had previously been his regular fireman, briefly took over the driving role before he too disappeared. By chance, a young man named John Snell arrived at much the same time and swiftly found himself parachuted into the jobs which the Joneses had just abandoned. Snell, at this time was about 19 years old and had recently left Bryanston School. He had written to Tom in March 1951 to ask whether he might come and work on the railway for the summer before going up to Oxford to read PPE. A triangular exchange of letters had followed between Tom; a London lawyer named Emley who described himself as Snell's 'quasi- guardian' and Talyllyn committee man James Boyd, whom Emley appears to have consulted about the railway's *bona fides*.[40] Initially, Emley insisted that Snell needed 'to take other work' – specifically, teaching in a prep school – after Whitsun, but this plan soon evaporated. Instead, Snell found himself taken on by the Talyllyn, initially as an assistant to Curwen, who was now the railway's Chief Mechanical Engineer, in the locomotive shed at Pendre.[41] Before long, he would find himself acting as Dai Jones' fireman and when Dai disappeared, he took on the role of driver.[42] Adept at making himself useful, Snell went onto the railway's pay-roll and continued to work for the Talyllyn throughout the entire summer and lodge, like Tom, at the Dolgoch Falls Hotel. It paved the way for a friendship that lasted for the rest of Tom's life.

Tom did not have an entirely easy time in his new role. He recalls in *Landscape with Figures*, that over the 1951 season, he was 'not unaware' of the resentment to which the authority that he exercised in his role of the railway's general manager gave rise among other Talyllyn supporters who made no secret of their injured feelings. They were, he says, 'saying in effect, "Rolt, Rolt, he's a canal man isn't he? What's he doing muscling in on our pitch and what does he know about railways anyway?"'[43] An article by Robert Humm suggests that the suspicion with which some of the working committee tended to view Tom's early interest in the Talyllyn and activity in the Preservation Society's early days was largely factional in origin. The Birmingham based members, Humm explains, already knew one another from attending the same meetings and rail tours and were used to acting in concert. Although an appetite for 'hands-on work' ensured that Tom and Curwen became well-known in Towyn, the 'higher management of the line', Humm claims, 'was [...] entirely a Birmingham matter'.[44] If Tom's name was known to the Birmingham-based Talyllyn enthusiasts, it would have been through his books, which up to this time had dealt with waterways rather than railways. Yet that detail hardly accounts for the resentment and suspicion which – if he is to be believed – Tom's role within Talyllyn circles appears to have given rise. Tom would claim to have made little of it, and dismisses it – together with Aickman's 'final purge of the heretics' in the IWA – out of hand. Nevertheless, the jealousy may have made a deeper impression upon him than he cared to admit.

On 23 October 1950, about a fortnight after the meeting in Birmingham's Imperial Hotel which had seen the formation of the working committee, two new members were co-opted to its ranks. One, James Boyd, came from a textile manufacturing family and in time became a well-known railway historian. The other, Birmingham industrialist John Wilkins, was unusually well-placed to offer practical advice and assistance on the Talyllyn because he was a co-owner of the Fairbourne Railway, a concern whose history may, in the early 1950s, have appeared to be extremely pertinent to the Talyllyn's new management.

Starting its life as a horse-drawn tramway used in the building of Fairbourne Village in the 1890s, in 1916 the Fairbourne concern had been re-born as a 15-inch gauge steam railway intended to develop the area's tourist appeal. Before it closed in 1940, it had rather a chequered career, which included an enterprising experiment with dual gauge track introduced to accommodate the impulse purchase of an 18-inch gauge locomotive.[45] One of the moving spirits behind its 1947 re-opening, Wilkins' experience may well have suggested to the Talyllyn's Birmingham supporters that he was someone with useful expertise to contribute to their venture. Sure enough, around November 1950 a special train ran with Wilkins on board, together, 'probably' with 'someone from the Fairbourne together with some [TR] committee members' for the purpose of conducting what amounted to a track survey. The fact that on their return from Abergynolwyn, they derailed 'somewhere east of Brynglas' probably gave the party all the information that they needed concerning the Talyllyn track.[46] Tom does not mention being part of this expedition although from his previous walks along the line he had, by this time, intimate first-hand knowledge of its condition. On 11 December 1950, a meeting of the Talyllyn Railway working committee heard a report from Wilkins in which he explained that the track was in a 'very poor state' with rotten sleepers and 'about 50% of the rails [....] unusable'. Wilkins' remit evidently extended to the locomotives and he informed the committee that *Talyllyn* was a 'complete write-off' while *Dolgoch*'s boiler required inspection.[47] That the committee should solicit a second opinion before spending money on the permanent way and motive power was not unreasonable, provided that the members went about it tactfully. If Tom's recollection of events in *Landscape with Figures* is to be believed, tact was in short supply. At the time, he claims, he brushed the slight aside being 'much too fully occupied with practical problems' to concern himself with 'the more unpleasant quirks of human nature'.[48] For what it is worth, nowhere in either *Landscape with Figures* or *Railway Adventure* does he mention Wilkins by name.

Although he does not labour the point, *Railway Adventure* gives some strong if tacit sense of the line's capacity to reconcile his interest in engineering with his feeling for the natural world in the same way that he had experienced years previously aboard *Cressy*. In essence, it found its place unassumingly within the existing landscape. The locomotives *Talyllyn* and *Dolgoch* had after all been a background presence in the lives of the people of Towyn – both at the Wharf and at Pendre – and at Rhydyronen, Brynglas, Dolgoch and Abergynolwyn, not to mention frequenters of Pendre, Hendre, Fach Goch, and Cynfal for the past eighty-five years. *Talyllyn,* appeared admittedly to be beyond repair, but *Dolgoch,* known more or less affectionately as the 'old lady,' continued to provide doughty if idiosyncratic service. Admittedly, her delicate patent valve gear often required remedial work by Tom or Curwen, and from time to time unforeseeable disasters like a piece of hard scale blocking a valve or a failed experiment with burning slack coal's resulting in her getting stranded near Brynglas occurred but, Tom asserts, for the most part, the railway maintained its service 'with exemplary punctuality'.[49]

He was perhaps not always the easiest of colleagues. Following the appearance of *Railway Adventure* in 1953, his first-person association with the railway tends to give the impression of his being the dominant, if not sole, moving spirit during the early years of the Talyllyn's management by the Preservation Society. It tended to foster the idea – in his own mind as much as in the view of others – that he took a larger part in saving the line than is, strictly speaking, true. *Railway Adventure* skates over Trinder's role in the Preservation Society's formation, for instance, barely mentioning his courteous overtures to Sir Haydn Jones and making little of his deft chairmanship of the public meeting of 11 October 1950. In fact, Trinder's tact and good sense must have been of inestimable value in ensuring that the Talyllyn Railway enjoyed enduring amiable relations both with the townspeople of Towyn and the holidaymakers who paid to travel on it. Besides, Trinder was a long-standing friend of Tom's from the days when the Rolts had been accustomed to make Banbury *Cressy*'s regular winter mooring place. Chronicling his own activity in the enterprise, Tom seems strangely blind to the central part that Trinder played in it. If Trinder felt under-appreciated, it may explain his decision to relinquish his chairmanship of the Talyllyn Railway Preservation Society in 1954 – its fourth year of operation – on the grounds that he felt 'unable to carry on'. At the time of his resignation, the Preservation Society made both him and Tom Vice-Presidents.[50]

In July 1951, Tom asked Philip Unwin to send copies of *Green and Silver* and *Inland Waterways of England* to Angela who, he explained, was 'travelling with a circus'.[51] Since in *Landscape with Canals* Tom leaves us in no doubt that Angela's departure from Tooley's Banbury boatyard in her £10 car had marked the effective end of their marriage and added that it would be twenty years before he saw her again, the terms in which he couches his request seem rather surprising.[52] Perhaps he did not care to admit to his publisher that his marriage had foundered; besides, he was also acutely conscious of his deepening fondness for Sonia.

Although the stages by which their relationship developed are far from clear, Tom stayed in contact with her while he was managing the railway and would later recall that they exchanged letters 'almost daily' throughout the summer of 1951. Hers, 'hastily scribbled […] in the small cabin of *Warwick*, arrived from canal side places that he knew well: 'Sutton Stop, Sampson Road, Warwick Two, Buckby, "Maffers", Bulls Bridge [and] Limehouse Dock'. From his lodging in the village of Dolgoch, he would reply to her by candlelight after the day's water supply to the local hydro-electric plant had run out. In September, Sonia told Tom that she had conclusively decided to 'leave the boats' and planned 'to lodge with friends […] in Braunston until she could decide what next to do'.[53] To maintain regular correspondence in circumstances which sound less than conducive to letter writing says much for their commitment to one another. At the same time, it is possible that their decision to spend the rest of their lives together may have seemed more obvious to their close friends and the people among whom they lived and worked than it appeared to either of them at the time.

If *Railway Adventure* gives no suggestion of the marital cross-currents that undercut Tom's life at the time when he was overseeing the railway from Wharf Station, they did not entirely escape the attention of his Talyllyn colleagues. Snell, for instance, clearly had some idea of what was afoot. 'Tom's personal life was changing in 1951 as radically as his professional life, because he was going through the process of [divorce]' he wrote years afterwards. Once the railway work of the day had finished, the volunteers had a chance to 'unwind and talk of other things'.[54] Observant and discreet, Snell followed their conversation closely, ready to contribute but with a sound instinct about when to keep quiet. His remarks about the law's insistence on a 'waiting period before a divorce was confirmed' and the necessity of avoiding contact with 'Another Woman' during this time suggest that among friends, Tom was open on the subject of the breakdown of his marriage to Angela. Sonia's name may also have entered the conversation and Snell's remark about never seeing her during the 1951 season suggests that he may not have been unaware of her existence. A number of Tom's friends from the VSCC, many of whom he had not seen for a considerable time, came to visit him in Wales. Affable, they soon drew Snell into their circle and John Morley took him for a thrilling drive over the Bwlch y Oerddrws mountain pass in his open topped Packard. Incidentally, Morley not only possessed an early Alvis – like Tom – but also the Maserati which Sir John Bowen drove in his last, fatal race.[55]

Tom's mother visited and Snell remembers her staying at Dolgoch and joining Tom, David Curwen and himself to climb Cader Idris. Despite her age, Mima was an energetic hill-walker and while Curwen and Snell were content to ascend as far as Llyn Cau – the lake in its glacial cwm that lies some distance short of the summit – mother and son 'shot on ahead' in ascent.[56] While it is, admittedly, no more than speculation, it seems not impossible that their speed may have owed something to a mutual concern that any conversation touching upon Tom's marital arrangements should take place in relative confidence.

Sometime before the railway's initial holiday season's running under Preservation Society management drew to its close, Tom wrote Sonia to suggest that the two of them should spend a week together 'on neutral territory'.[57] It might, he thought, help them to decide what steps they should next take. If his approach sounds cautious, to a couple who had both seen their previous marriages break down, caution probably commended itself.

Chapter Twelve

Towards a Settled Life

Tom admits to sighing with the relief at the sight of the final train of the Talyllyn's 1951 season drawing into Wharf station. Having allowed a few days for a final pre-winter clear up, he drove to Shrewsbury to meet Sonia and take her to stay at an inn somewhere between Clun in Shropshire and Newtown in what then was known as Montgomeryshire. In the light of his knowledge of the more southerly Welsh Marches, his claim that the nearby country held no memories and no associations for either of them feels rather tenuous. It is, for instance, difficult to believe that their time in the Kerry Hills stirred no memories of his Cusop childhood. Defiant, Tom insists that on some occasions it behoves an autobiographer to be 'reticent'. What is clear is that over a week's walking, talking and shared enjoyment of the 'superlative' meals which their landlady provided every evening, the couple realised that they were much in love. Nevertheless, their mutual lack of funds precluded any question of their 'setting up house together, even in the most modest fashion' so, at the end of their holiday, Tom drove back to Stanley Pontlarge and Sonia returned to her friends in Braunston.[1]

But Mel Russell-Cooke – she who had joined Tom for *Cressy*'s final journey – evidently had a shrewd understanding of how things stood between the couple and made the tactful suggestion that he and Sonia might care to look after her Oxfordshire home, over the winter of 1951 while she went to Ireland with Tom's painter-cousin David. What was more, Mel had no objection to their using her house for what Tom euphemistically terms 'the dreary charade [...] necessary to provide the required evidence for divorce'; in other words, a carefully staged act of adultery which one of Tom's VSCC friends 'nobly' agreed to witness. For Tom and Sonia, it was not an easy time. To judge from *Landscape with Figures*, they both struggled to adapt to leading a leisurely life at close quarters in Mel's elegant, not to say luxurious, house without the routines that their respective railway or canal work supplied. Writing in retrospect, Tom asserts that both of them experienced acute misgivings on account of George and Angela.[2] To fill the days, they explored the surrounding country or went shopping in Witney or Oxford. In the evenings, Tom worked on his current book-in-progress.

A celebration of the small, obscure and sometimes eccentric railways that he enjoyed, Tom intended to give it the title of 'Railway Byeways' and conceived it as a collaboration between himself and his Talyllyn Railway colleague Patrick Whitehouse. He would write the text and Whitehouse, a photographer to whom he attributed 'an eye for pictorial quality [...] as well as for the railway interest' would provide the pictures.[3] For the finished work, Tom asked Allen & Unwin for payment of a substantial 15% royalty, intending that he should keep 10% and give Whitehouse the 5% balance. That he should seek to drive so hard a bargain suggests that he was increasingly worried about his finances.

Philip Unwin was not unsympathetic, but he would not and could not sanction payment of so generous a royalty on a relatively unassuming publication. Priced at 15s, the book could anticipate good sales; priced any higher, he foresaw that it might well lose out. If Tom chose to publish with a firm who were willing to offer more generous terms, he wrote, Allen & Unwin would 'stand aside' – that is, release him from his contract – albeit with regret.[4] The argument continued through February and into March 1952 when it emerged that Constable were ready to show what Tom calls 'co-operation over the railway book'. They were, in other words, keen to publish it on his terms, and he 'could not do other' than accept their offer.[5] In the course of writing, he changed the title to *Lines of Character*.

When he went to Ireland in spring 1952 to gather material for the book's Irish section, Sonia joined him. Their respective divorces had not been finalised, but it would be what Tom calls 'a sort of unofficial honeymoon' since by this time the couple considered that their relationship was stable, secure and lasting.[6] Among the railways that Tom wished to visit, the Tralee & Dingle Light Railway held a special place. Passenger carriage on the line had, in fact, ceased in 1939 – its Castlegregory Branch closed the same year – although a daily goods service survived until 1947. At the time of Tom's and Sonia's visit, the railway's sole service was a monthly cattle train that ran to Dingle for the market.[7] Such a limited timetable restricted their options for travel, but Tom found a likely date and Córas Iompair Éireann – one of Ireland's statutory transport authorities – provided Sonia and himself with footplate passes.

Tom would describe their trip as the 'most unforgettable railway journey' that he had ever made or was ever likely to make. The feature that he highlights in *Lines of Character* include the line's extreme curves and stiff gradients; the spectacular scenery through which it passed; the viaducts at Curraduff and Lispole, both scenes of major accidents, respectively in 1893 and 1907; and the lack of coaling facilities on the line which meant that fuel occupied most of the space inside the locomotives' small cabs. Somehow, he ensconced himself behind the driver in the cab of the lead engine, while Sonia gamely secreted herself in the corresponding space on the train engine.[8] While Tom's driver evidently had years of familiarity with the terrain and the locomotives, Sonia found herself at the mercy of a more impetuous individual who was given to releasing his brake well before he had reached the bottom of a descent in order to build up momentum with which to charge the next hill. But they had a wonderful journey. From Tralee, they passed through Blennerville, came over the Glenagalt bank with its 1 in 30 gradient – Tom's driver maintained that it was 1 in 25 in places – crossed the Lispole viaduct where, for safety reasons, the pilot engine had to be uncoupled to 'run light over the bridge [...] to await the rest of the train' and, five miles later, reached the end of the line at Dingle.[9] The westernmost town in Europe – it was, in Tom's words, 'a kind of railway *ultima Thule*' – when he planned *Lines of Character*, he decided that their arrival there made an appropriate conclusion to the book. Nevertheless, it was not actually the end of his and Sonia's Tralee and Dingle Light Railway experience.[10]

Taking up the tale in *Landscape with Figures*, he explains that the Cattle Fair was over by noon and the returning train for Tralee left Dingle Station at 1 p.m. On one of the steeper ascents, Tom glanced back to see that one of the loaded cattle trucks had derailed. Brisk discussion among drivers, firemen and guard ensued. Tom and Sonia, it was agreed, would travel on to Castlegregory with the foremost portion of the train, while the engine crews summoned a breakdown gang to attend to the derailed wagon and arranged for the train that was following them from Dingle to be held at Anascaul.

When they reached Castlegregory, avid to see how the drama would end, Tom and Sonia asked to return to the scene of the disaster. On their arrival, they found the errant wagon back on the rails once more. Their engine, her cab now very crowded, propelled it back to Anascaul station where it was shunted in with the other wagons and, for the second time in the day, they set off towards Tralee. As it happened, three plate-layers employed on the line were still at the site of the derailment which was fortunate because 'right in front of their noses' another wagon managed to derail itself – the second of the day..[11] When eventually their journey resumed, their driver was anxious about whether the train would get past the Lee estuary before the incoming tide flooded the track, and Tom would remember him shaking his head and checking his turnip watch by the light of the firebox as they approached the coast. Evidently a well-recognised hazard among Tralee and Dingle drivers, some of them tended to wait cautiously for the tide to fall, which might take anything up to three hours, before proceeding along the estuarial stretch of line, while others rushed ahead, taking the chance of the waves quenching their fires, leaving themselves and their passengers marooned.[12] Tom and Sonia's train passed the estuary without incident and, having made their farewells to the engine crews and watched the drovers herd off the cattle, they retired to their hotel for a substantial 'tea'.

Lines of Character, which discusses distinctive railways of England, Wales, Scotland and the Isle of Man as well as those of Ireland, came out in November

1952 priced at 21s and received some appreciative notices in the press. H.L. Howarth who reviewed it for the *Birmingham Daily Gazette* described it as a 'welcome book' and it received flattering attention from the *Western Mail* and the *Crewe Chronicle*.[13] But Philip Unwin's assessment of its chances proved well-founded. For all its *joie de vivre* and gusto, *Lines of Character*'s publication unfortunately coincided with an immense number of other railway books and, to adopt Tom's metaphor, in a flooded market, it 'sank without trace'.[14]

Soon after their return from Ireland, Tom took Sonia to Towyn for the Talyllyn's second season of running under the Preservation Society's aegis. The committee had asked him to 'take charge' of the railway again, but David Curwen had business commitments in Devizes which precluded him from looking after the engineering side of things, so Tom foresaw that his role was likely to include locomotive repair and maintenance. Because he could not be in the workshops at Pendre while managing Wharf station, he proposed that Sonia take on the public facing role of managing bookings and ticket sales. Since she was happy with the prospect, it seemed the perfect way ahead, but some of the TRPS committee objected to the arrangement on the prudish – or perhaps mischievous – ground that, since Tom's and Sonia's respective divorces had not yet been made absolute, the couple were 'living in sin'. Extraordinarily, the dispute apparently took up 'many weeks' which cannot have been a comfortable time for anyone involved in the railway's management. Matters were at last resolved by Edward Thomas's observing, with all the gravitas of an 'elderly chapel-going Welshman' that he thought Tom's plan 'excellent' and that the private lives of the volunteers were 'nothing whatever to do with the Railway Company'. Taking encouragement from his words, Tom and Sonia hired a caravan, stayed for the summer at Dolgoch and took charge respectively of Pendre and Wharf Stations.[15]

One August evening at Wharf Station, while he awaited the arrival of the day's last train, Tom found himself greeted by a clergyman who offered his service as a volunteer for a week. Knowing that the railway was short of a guard, Tom at once recruited him to fill the post. His name was Wilbert Awdry.[16] He soon learned that the work of a Talyllyn Guard encompassed that of travelling booking clerk; in other words, he had the task of issuing tickets and giving change to passengers who joined the train at any of the stations or halts along the line between Towyn and Abergynolwyn. By his own admission, he had an uncertain command of mental arithmetic which often resulted in his giving the passenger too much change, leading him to make up the shortfall to the railway from his own pocket. He was also apt to forget about the brake in his van and either omitted to put it on at stations, which could make it difficult to start a heavy 'up' train with tight couplings or, having screwed it down, neglected to ease it off which could cause the locomotive to slip. His worst mishap came one evening when he gave driver Bill Oliver the 'Right Away' signal for departure from Abergynolwyn before he had checked that Mrs Davies, who had been selling soft drinks and sandwiches all day to passengers and passers-by, was safely aboard the train. Only once they had set off and he heard a yell and saw her, flushed and frantic on the platform, did he realise what had happened. Against the clatter of the locomotive, blowing his guard's whistle was vain endeavour but by screwing his brake hard down so that the train palpably dragged he managed to catch the attention of the engine crew and Bill Oliver 'propelled the train back' to collect the distraught Mrs Davies. That she was his mother-in-law may have heightened the awkwardness of the situation.

For Tom, the encounter with Rev. Wilbert Awdry marked the beginning of a powerful friendship. While Awdry's church commitments left him with little free time – when he started his first stint as a Talyllyn volunteer, he was Rural Dean of Bourn in the diocese of Ely – railways were probably never far from his thinking.[17] Of the books that make up his 'Railway Series', some seven were in print by 1953, the year after his initial, eventful week as a Talyllyn guard. Clergy duties and authorship were both time-consuming. Awdry's wife Margaret and their children – son Christopher and daughters Veronica and Hilary – took a more-or-less patient view of his rail-passion, but the Talyllyn did not possess them in the same way that it did Wilbert. Brian Sibley, his biographer, remarks that the family's 1952 Towyn holiday confirmed Margaret Awdry's 'worst fears'; while her husband absorbed himself happily in railway matters, she had to take sole charge of the children.[18]

But Tom was a true kindred spirit. Long ago, he had written a number of stories about garrulous locomotives who were quick to steal a march on one another. They were never published and whether he ever told Wilbert about them is unknown, but it is quite likely that by the 1950s he knew about Wilbert's 'Railway Series' of books because Ray Aickman, also known as Edith Ray Gregorson, was the literary agent who sold them in 1948 to Leicester-based publisher Edmund Ward.[19] Tom whose knowledge of railway history enabled him to recognise the incidents behind Wilbert's fiction, may have suggested that Wilbert might care to incorporate episodes from recent Talyllyn experience within his narratives.[20] While it did not come about immediately, when *Four Little Engines* appeared in 1955, Mrs Davies – assuming that she read it – would find the occasion when she got left behind immortalised in the story of 'Peter Sam and the Refreshment Lady.'[21]

Tom's practice was to spend his summers busy in manual outdoor work and to reckon to write during the cold and dark of winter. Necessarily, this routine relied upon his having a suitable winter base and as the days grew shorter and colder, the Dolgoch caravan's disadvantages – standing in the shadow of the hills it was dark and chilly – became increasingly marked. Of Sonia's and his moving into Stanley Pontlarge, which might have seemed logical, there was no question. Although he does not admit it outright, it seems likely that Mima could not accept that his marriage to Angela had ended and that he had fallen in love with another woman who was herself married. Born in 1880, Mima belonged to a generation which tended to view marital breakdown as a species of scandal. She declined to invite the couple to share her home and Tom doubts that he would have accepted any such invitation, even if it had been forthcoming. Mother and son were evidently at loggerheads. Fortunately, Sonia had an aunt based in Roehampton who 'had acted *in loco parentis*' towards her following her mother's death. Having made contact with this lady – she was 'a sterling character', Tom reports – Sonia and he stayed with her for the winter and their son, Richard, was born at her home in January 1953.[22] A second son, Timothy – known as Tim - was born to the couple in 1955.

Mindful of the need to provide for his family, around this time Tom wrote a book about his two years of managing the Talyllyn Railway. For *Railway Adventure*, as he called it, he entertained high hopes. Not only did he envisage it as 'a railway equivalent of *Narrow Boat*' but he designed a jacket for it in the same style as that of the earlier book and persuaded John Betjeman to write its Foreword. When it failed to achieve the sales for which he had hoped, he was understandably disheartened.[23] But its poor showing was perhaps not entirely surprising. Unlike *Narrow Boat*, a journey narrative that also surveys the condition of England, *Railway Adventure*'s remit – the preservation of a railway in mid-Wales – is narrow; its lineside vision was parochial and its appeal – though great among rail enthusiasts – was likely to be limited with the public at large. More seriously from a sales point of view was the book's tendency – which did not entirely escape its reviewers – to present the holidaymaking public who travelled on the Talyllyn in less than flattering terms. For travellers who mispronounced Welsh place-names or struggled to grasp the intricacies of the timetable, Tom shows little patience. To visitors intent on sharing fond railway memories with busy staff, or apt to wander off when they reached Dolgoch or Abergynolwyn, assuming that the train would wait for them indefinitely rather than run to timetable, he gives short shrift. He is, however, kinder to them than he is to parties of school children whom he dismisses as 'demonic little creatures' and 'hooligans'.[24] While, broadly speaking, he liked *Railway Adventure*, railway writer Canon Roger Lloyd observed that Tom's writing was often 'provoking' and remarked that 'an unduly large proportion of the human race fails to please him'. Canon Lloyd found his sniping about the recent railway nationalisation particularly objectionable, especially since the Railway Executive had made the two ex-Corris locomotives available to the Talyllyn on most generous terms.[25]

Over the spring, summer and early autumn of 1953, Tom, Sonia and the infant Richard led a wandering life. Feeling that they had trespassed for rather too long on the goodwill of Sonia's aunt, they decamped in March 1953 to Devon, where Sonia's friends Tudor and Meriol Trevor offered them the use of the top floor of their house at Aller Park near Bideford. By mid-summer, they had moved to Shropshire, where Tom took a lease on Laurel Cottage in Clun.[26] That he was not entirely happy with the Talyllyn Railway's day-to-day organisation is evident from his surviving letters to Bill Harvey, shed master at Norwich Station, who had come to Towyn as a volunteer in 1952. In a letter to Harvey of 30 December 1952, he castigates his Talyllyn colleagues' lax approach to the railway's need for a steady cashflow. 'Even enthusiasts have got to live', he writes:

> and [...] the future of the railway cannot be assured until we are in a position to ensure that whoever runs the railway won't be out of pocket. The only answer is a far more vigorous publicity campaign to swell the Society's funds, but they have foregone my help with publicity and are being very casual about it. They seem to think I am concerned to feather my own nest...What is so exasperating is that some of these Birmingham blokes just can't be made to realise that any problem exists.

In May 1953, shortly before Sonia, Richard and he moved to Clun, he states that he had seriously considered settling in Towyn, 'but', he discloses:

> some of the B'ham boys would never give me that confidence, goodwill and wholehearted support which alone would have justified my taking such a step.[27]

By this time, the Preservation Society had appointed Ken Marrian as general manager.[28] Tom was apt to take a critical view of some of his decisions. On a hurried visit to Towyn in July 1953, he had been less than impressed by Marrian's response when *Sir Haydn* had derailed. 'Marrian seems a nice enough chap', he told Harvey:

> but he seems to leave the most vital department – Pendre – to muddle along any old way. I know he's not a loco man, but as GM he ought to exercise some supervision over what goes on. No. 3 coming off the road with a full afternoon train should, I would have thought, be treated as a first class crisis with 'all hands to the jacks' but Marrian never left the office and did nothing to help bar ringing up for transport to bring the passengers back.[29]

It was not an easy season. Charles Uren, the railway's chief mechanical engineer of the time, recounted ongoing struggles with *Dolgoch*'s cracked firebox, *Sir Haydn*'s leaky tubes and *Edward Thomas*'s recurrent valve trouble.[30] If Tom was worried, he may have had good reason. Allan Garraway who was, like Harvey, a professional railwayman – specifically, assistant to the Motive Power Superintendent for BR's Eastern region – shared Tom's misgivings and wrote bluntly to the effect that while Marrian was an agreeable companion and competent booking clerk, he was less than adequate in the role of general manager.[31] No reply from Tom survives, but the gulf that existed between him and the Birmingham element of the Talyllyn is unmistakeable. As the 1950s progressed, Garraway played a leading part in restoring the Ffestiniog Railway. But with increased Talyllyn experience, Ken Marrian seems to have found his footing. Despite health problems, he remained in the general manager's post until his death in January 1958 when Patrick Whitehouse wrote an affectionate 'Appreciation' in which he credited Marrian with a 'wonderful gift for making friends with everyone'.[32]

Although Tom and Sonia liked Clun, Laurel Cottage which they leased was a dark house. Built partially into a bank, its natural light was somewhat restricted and, worse, it had no space in which to park the Alvis.[33] When the chance to purchase it arose, these disadvantages were substantial enough to dissuade them from making any offer. Instead, they looked at other houses near the Welsh border, but the few that they could afford 'invariably' came with attendant problems. Besides, Tom knew that Stanley Pontlarge urgently needed repair and maintenance, its roof being in a decidedly dangerous condition, which Mima who thought only of her adored garden, had neither the means not the inclination to set in motion. Even if he found a suitable house in which Sonia and he could raise a family, the question of what to do about his mother and his old home remained pressing. Then, probably in late summer or early autumn 1953, Mima unexpectedly announced that she wished to see Tom, Sonia and Richard, her grandson, so they drove over from Clun to Stanley Pontlarge for a day's visit.[34]

About what passed between them, Tom says nothing, but it must at least have started some debate about their future accommodation arrangements. The outcome was that Mima agreed to make the old house at Stanley Pontlarge over to Tom by deed of gift while he, in return, undertook to build a 'dower house' for her in the orchard. Quite what swung her thinking is impossible to say. Even when the building works got underway, it cannot have been an easy time for anyone and *Landscape with Figures* makes no secret of the problems attendant upon wiring the ancient dwelling for electricity and equipping the new cottage with drainage. It was, Tom pronounces succinctly, and probably with a measure of understatement, 'a very difficult transition period for all concerned'.[35] The question of how to accommodate his and Sonia's accumulated books proved problematic until a resourceful Talyllyn volunteer who had already helped with wiring the old house brought a supply of wooden planks from Wales from which Tom constructed bookshelves.[36]

Amid the stress of the sheer upheaval of moving house and uncomfortably aware of the disappointing sales both of *Lines of Character* and *Railway Adventure*, he was beginning to question his ability to support his family by writing. Surprisingly, perhaps, his next book would be a work of fiction, albeit one which drew upon his knowledge of the industrial revolution which he would chart in his history of an imaginary Midlands town. *Winterstoke* – as he names both the place and the book – revisits ground that he had explored previously in 'Strange Vista,' but lacks anything resembling the romance of David Hancock and Anne Cray which gives the earlier story its core. *Winterstoke* does not concern itself with the dynamics of relationships and provides no central characters with whose lives its readers can engage. Essentially a book of issues, with his gestures to the nuclear research establishment at 'Summersend Hill' – Tom always enjoys sinister place names - it moves inexorably towards a conclusion which, although the scenes of mass panic that surround the pandemic in 'Strange Vista' play no part in it, promises to be apocalyptic. Having grown from a modest settlement by a monastery to a thrusting centre of manufacture, by the mid-twentieth century – the time of the book's conception – Winterstoke comes increasingly to resemble a town which has ' lost its soul'. Even if its 'wheels spin ... traffic roars ... and long expresses slide into the central station' and its inhabitants 'crowd pavements, chainstores and market more densely than they ever did before', in Tom's vision, the bustle has no more spark or significance than 'the automatic muscular contraction in the body of a chicken after its head has been struck off'.[37] While he leaves us, as we read, in no doubt about the threat that the Summersend laboratories present, he casually slips in the information that the farmer, William Winter, long-term descendant of the first inhabitants of the place, has bequeathed some property to the Church. It offers some slender hope for the future and the book concludes with the statement that 'after four hundred years the Cistercians have come back to Winterstoke', albeit to 'confront another wilderness'.

Winterstoke came out in the summer of 1954. Reviewers on the whole tended to like it. 'E.J.L.' of the *Tewkesbury Register* commended it earnestly 'to all who are interested in social and economic history' while in the *Birmingham Post* Edmund Vale called it a *tour de force* and nominated it as one of his 'Books of the Year'.[38] But to the book-buying public at large Tom's fiction in the guise of history made little appeal and *Winterstoke* never earned out the modest £100 advance on royalties that Constables had paid him. Yet if its admirers were few in number, some of them wielded significant influence. John Guest, literary advisor to the publisher Longmans, liked it enough to write to Tom expressing his admiration and proposing that they met. It would mark the start of a fruitful relationship.[39] Much more recently, in 2015, Faber reissued *Winterstoke* as part of the 'Faber Finds' series, 'dedicated to bringing great writing back into print' which suggests that even if it has never been a bestseller, it will always have its champions.[40]

With hindsight, Tom came to marvel at the publisher Constable's eagerness 'to accept any book' that he offered them and it was he, rather than the firm, who ended their relationship. Constables, he observed 'failed to make a success' of any of the five of his books that they had published, 'on a very wide range of subjects'.[41] Unwittingly perhaps, in this remark he touches upon what may have been the nub of the problem. Mercurial himself, and ardent in his pursuit of diverse interests, it seems likely that neither he nor his publishers could come to terms with the conservative tastes of the book-buying public. Popular writers tend to owe their success to loyal devotees – essentially a fan club – who are confident of knowing what to expect from their work. With L.T.C. Rolt, it must often have seemed as if there was no knowing what arcane subject he might pursue next. From ghost stories to vintage cars, from idiosyncratic railways in Wales to the study of the rise and fall of an imaginary town – to readers with well-developed tastes for specific genres, his sheer range must have been a puzzle. Railway enthusiasts' taste for ghost stories, after all, cannot be taken for granted; nor, indeed, will avid readers of supernatural fiction necessarily wish to read about life on the canals, the history of the motor car or the plight of the artisan faced with the increasing mechanisation of traditional craft industries. By casting his net so wide, Tom unwittingly precluded himself from building up a regular following. Although Constable were ready to keep faith with him, he doubted their ability to sell his titles and, to the disappointment of company chairman Ralph Arnold, broke their option on his future books.

Fate, at this point, turned in his favour. The VSCC asked him to act as clerk of the course and secretary for a vintage car rally scheduled for the late summer of 1954. The event owed its inspiration to the British Travel and Holidays Association which, building perhaps on the excitement that the 1951 Festival of Britain had generated, wished to promote Britain as a destination for American visitors. Taking on the dual role would play both to Tom's skills as an organiser and to his appreciation of the specific demands of the vehicles in terms of fuel and maintenance, while at the same time promising him a welcome and timely cash injection.[42]

His experience of running the Anglo-American Car Rally occupies a substantial chapter of *Landscape with Figures*. It took place over seven days: 4–11 September 1954. The route on which Tom and British Travel and Holidays Association representative Mr McNab agreed would take the vehicles from Edinburgh Castle to Chichester, by way of Alnwick; Harrogate; Boston; Cambridge; Leamington and Cheltenham, with concluding vehicle tests at the Goodwood circuit in front of a packed grandstand. Each day's schedule included a generous 'lunch stop' giving the participants a chance to explore the Borders village of Dirleton; the cities of Durham, York and Ely; Sulgrave Manor; Stratford-upon-Avon (to be followed by a tea stop at Prescott Hill); and, on the penultimate day, Savernake, followed by tea in Winchester. Clearly, the Travel and Holidays Association were keen that the American visitors should have ample chance to enjoy historic sites and varied terrain. While the rally was in progress, a number of VSCC members accompanied it in their own vintage vehicles to act as stewards and marshals, Tom and Sonia among them in the Alvis, having left Richard at Sonia's aunt's Roehampton home in the charge of a nanny.

The event aroused great public interest. 'American Old Crocks coming' announced one North of England paper in excitement even before the American drivers and their cars had embarked for their transatlantic crossing.[43] Having docked at Liverpool on 28 August 1954, shepherded by the Rolts and VSCC member Arthur Jeddere-Fisher and his wife Marcia, they set off for Edinburgh, catching much attention along the way. At Grange-over-Sands, for instance, the chairman of the district council came to welcome the American drivers in his vintage Invicta which at once – perhaps to his surprise – attracted numerous offers of purchase.[44] Apart from Tom's embarrassment upon discovering that the Scottish Tourist Board had housed the visitors in decidedly sub-standard Edinburgh accommodation – 'reading the Riot Act' in the relevant quarter soon remedied the situation – the rally went ahead in genial spirit.[45] Its official start at Edinburgh Castle came in for coverage by the BBC Light Programme with commentary from Robin Richards and as the event got under way, it drew much attention.[46] *Daily News* reporter Peter Vane travelled in American team captain H. Austin Clark Jr's 1961 Pierce-Arrow Raceabout for the Alnwick-Harrogate stage, precariously balancing his sixteen stone bulk on the bucket seat to observe cheering crowds all along the route.[47]

As a result of Tom's lobbying, the field included a Stanley steam car which gave the spectators more delight and the rally officials more worry than any other participating vehicle. Its driver and owner Paul Tusek, postmaster of Power Point, Ohio, had not yet entirely assimilated himself to steam cars' characteristics and he was apt to overlook the need for frequent refilling of the Stanley's water tank.[48] On the way to Edinburgh from Liverpool before the start of the rally proper, Tusek, his car and his passenger-cum-travelling-mechanic – an affable New Englander named Ed Battison – vanished somewhere north of Garstang, apparently without trace. It emerged that their water had sunk alarmingly low and they had had to refill the Stanley's tank at a remote farm.[49] The same disaster almost struck again en route to the Scots border when Tusek forgot or omitted to heed Tom's instruction to refill the tank at Moffat. He drove through the town heading north, leaving Tom to intercept him in the Alvis and shepherd him back to Moffat's public convenience, where the 'tank-filling operation was smoothly and swiftly carried out' by means of a length of hose attached to a tap, in front of a large and no doubt curious crowd.[50] A few days later, trouble struck again between Alnwick and Harrogate, this time with the Stanley's vaporizer; carbon from unfamiliar fuel (supplied *gratis* by Shell) had blocked the tube which had to be drilled out.[51] The Stanley's next misadventure reached the pages of the *Daily News*, which under the headline 'PUFF, PUFF Then old Stan Blows his top,' gleefully reported that the 'gallant American old-timer' had blown 'the packing on a water pump' on the way to Cambridge. On this occasion, a remedy for the damage was effected relatively easily, 'with the expert application of a couple of apparently prehistoric spanners'.[52]

A more serious setback occurred at South Harting Hill in West Sussex, on the rally's penultimate day and tantalisingly close to the finish. Having achieved one triumphant ascent, Tusek attempted the climb again, only for the Stanley to falter near the summit, pause and then stop with 'a hissing sigh of steam.'[53] To Tom, it was evident that something had gone badly amiss with the car's valve gear and he arranged for a lorry to take it into the Chichester garage where the rally vehicles were to be accommodated overnight. Having called on the assistance of Talyllyn stalwart David Curwen, throughout the night they tried conclusions with the Stanley, but the sheer difficulty of removing the steam chest cover – it was a substantial cast iron plate which concealed the defective valve – frustrated their endeavours. When eventually they managed, with much trouble, to detach it, it became evident that corrosion and rust had taken their toll upon its threads and that it would never be fit for purpose again. In the short time available, there could be no question of finding and fitting a replacement cover. Tom and Curwen applied sealant and warned Tusek that he must attempt nothing more on the morrow than one or two 'slow demonstration laps'.[54] In the event, he pulled out of the closing part of the rally.[55]

On the final day, the competitors completed their reversing, parking and bending tests, together with a fifteen lap trial of stamina before making a final two exhibition circuits and concluding with an end-of-rally party in Goodwood House. In terms of forging friendships and giving the nation at large a fine show, the rally was a huge success. Insofar as it had been competitive, the home team won by what Tom considered an 'embarrassingly large margin', but the Americans got their revenge at a follow-up rally in the US two years later where they trounced their British rivals.[56] Torn between relief that it was over, and the 'empty feeling of anti-climax' that came in the wake of the excitement that it had generated, exhausted, he returned with Sonia to Stanley Pontlarge – now their settled home – to resume work on his current book-in-progress.

Chapter Thirteen

Second Breakthrough

For years, Tom had regarded himself as 'something of a student of railway accidents'.[1] To anyone who had read his short story 'The Garside Fell Disaster' his interest would hardly seem surprising, and at some point in about 1953–4, Richard Hough (1922–1999) of the Bodley Head – to whom Charles Hadfield's friend Frank Eyre had introduced him – suggested that he might care to write a book on the subject.[2] It is possible, although no more than conjecture, that Hough's suggestion sprang from the Harrow and Wealdstone railway disaster of 8 October 1952. An appalling accident, in which a London-bound express from Perth had crashed into a local service from Tring at Harrow and Wealdstone station before a north-bound train struck the wreckage, it resulted in some hundred and twelve deaths, and shocked the nation.[3] Many of the fatalities, it emerged, were Euston-based rail staff.[4]

Tom recognised that he was well-placed to undertake the necessary research. Only recently, the Talyllyn had come in for a formal inspection by Colonel Denis McMullen of HM Railways Inspectorate. Despite some initial misgivings and Col. McMullen's calling for repairs to a couple of underbridges, Tom found the experience 'constructive and helpful' which suggests that it passed off in mutual goodwill.[5] The fact that McMullen was, by chance, an old Cheltonian may have encouraged Tom to ask him for permission to use the Railway Inspectorate's extensive archive of accident returns and reports. Certainly Col. G.R.S. Wilson, the Chief Inspecting Officer of the time, was willing not only to allow him access to the material, but also provided him with a research room.[6]

In the early days of the railways, intrepid travellers evinced an extraordinary want of respect for obvious dangers. Tom remarks, for instance, that by the time that he came to write *Red for Danger*, the specific need to enact byelaws forbidding passengers to travel on carriage roofs appeared laughable.[7] In the 1820s and 1830s, the peril had seemed less obvious. So indeed, did the self-evident necessity of a system of block working with the line divided into sections to be occupied by only one train at a time. Dr Dionysius Lardner was only one of a number of self-appointed experts to whom the Great Western Railway gave, for a time, usage of a dedicated train in which to roam freely over its tracks conducting their own research, to the peril of other rail users.[8] But as the network expanded and passenger numbers increased, matters of railway safety assumed increasing importance in the thinking of the travelling public.

The Railway Regulation Acts of 1840 and 1842 empowered the Board of Trade to appoint inspectors, charged, first, with surveying and reporting on new lines which could not be opened for public use without their approval and, second, with investigating accidents and reporting their findings. Generally officers of the Royal Engineers, the inspectors had a fair understanding of the challenges that operating and managing railways could present and, in Tom's view, they sought 'to steer a middle course' between defensive railway companies on the one hand and the anxious – and sometimes outraged – public on the other.[9] It was, perhaps, self-evident that in their work they should seek, 'to eliminate the possibilities of human error' as Tom expresses it, or, in the words of a recent Office of Rail and Road publicity video, 'turn tragedy into foresight'.[10]

The volume of paperwork which the Inspectorate's enquiries had generated over the years was immense. Early on in his research, Tom recognised that if he were to avoid drowning in primary source material, he had to devise a clear schema by which to organise his book. By chance, he owned an old edition of the *Railway Year Book* – a publication which flourished in the 1910s and 1920s. Not only did it tabulate accidents in chronological order, but it also, usefully, provided summary details of their causation under the heading 'Remarks'.[11] It enabled him to categorise his raw subject matter by topic so that he could devise a book which, chapter-by-chapter, would discuss and examine different species of disaster under such labels as 'Blow-ups and breakdowns'; bridge failures; and signalmen's and drivers' errors. While the Harrow and Wealdstone disaster – the worst peacetime rail crash in British History – may have been central to the book's inspiration, Tom decided that since it was still raw in people's memories, it would be 'inappropriate' to discuss it in his text.[12] Nevertheless, in *Red for Danger* he is clearly conscious of addressing a readership whose knowledge of railways was general rather than specialist – people in whom the recent accident had stirred some curiosity about railway working and procedures. His intention, he explained in the book's Preface, was not to sensationalise his subject matter, but rather to trace the Inspectorate's role in analysing each accident's cause and recommending specific safety measures, designed to avoid its recurrence. From Shakespeare's Hotspur, he draws the book's talismanic motto: 'Out of this nettle, danger, we pluck this flower, safety'.[13]

Rather than giving a history of lock and block signalling, braking systems and other railway safety features which would have been a valid, if dry way of approaching the subject, the book focusses on the individuals concerned in the railways' operation. From an accident at Wellingborough in 1898 when a luggage trolley ran down a slope onto the line into the path of the Manchester bound express causing a derailment in which seven people lost their lives, he concluded that:

> safety on the railway does not depend on enginemen and signalmen alone, but that railwaymen of every grade from porters to permanent ways staff carry a share of responsibility.[14]

That is not to say that any individual was held culpable for the disaster; in this instance, the Midland Railway Company admitted liability. In his report, the Inspector, Col. Yorke, recommended first that station platforms should either be level or slope away from the track rather than towards it and, second, that trolleys be fitted with brakes that remained on, unless or until the trolley was actually pushed. Nevertheless, by focussing his survey of railway accidents on people rather than procedures, Tom sets railway management and railway responsibility within a human sphere. On a narrative level, he not only made an extensive exploration of each accident that he discusses, but also enters into the thinking of the Inspectorate as they sought, in his words, 'to steer a middle course between recalcitrant railway companies on the one hand and an indignant or alarmed public on the other.'[15]

Critical to railway safety is the principle of 'lock, block and brake'. In essence it means, first, that points and signals should interlock so as to avoid any chance or risk of collision; second, that signalmen should be provided with block instruments so as to enable them to keep an 'absolute space interval between successive trains' and third, that both the driver and guard of a train should be equipped with an effective brake which they could apply and which would come on automatically in the event of a train becoming divided.[16] If the idea seems obvious, its adoption did not come about quickly. The Armagh rail accident of 1889 illustrated all too graphically the consequences of disregarding the 'lock, block and brake' precept and would lead to its becoming a mandatory requirement.

Rather than having a block system in place, the Newry and Armagh line operated on a time-interval system. It meant that the crew of a train leaving any of its stations could not know for certain that the way ahead of them was clear. On 12 June 1889, driving a relatively small 'four-coupled locomotive', the inexperienced Thomas McGrath had a train of fifteen carriages with 940 passengers on board. Having declined the offer of banking assistance, a long gradient defeated his locomotive and he and his colleague James Elliott – chief clerk to the railway's superintendent, who had travelled on the footplate with him – faced the choice either of waiting for the next train to push them up to the summit, or dividing the train and taking the front portion ahead to the next station before returning for the second portion. They opted to divide, with appalling consequences. Having uncoupled the rear half of the train and put stones as scotches under its wheels, the locomotive set off with the front part of the train. But as McGrath started, before gathering forward momentum, the locomotive initially rolled back, so as to apply slight but significant pressure to the detached carriages. On the steep gradient, they began to roll downhill, crushing and scattering the few stone scotches, straight into the path of an oncoming train. Although the guard tried to apply his brake, it proved useless.

'No-one' Tom remarks, thought of seizing one of the old rails which were lying at the lineside and thrusting it through the wheels of the runaway 'as a sprag,' although quite how easily it could have been achieved is open to question. In the event, ten carriages of the excursion train ran headlong into the next train on the line which had followed at the regulation twenty-minute interval. Some eighty passengers lost their lives in the ensuing collision, many of them children.

Analysing the causation of the Armagh disaster, Major-General Charles Scrope had some criticism of the Great Northern Railway of Ireland for 'assigning to their locomotives duties so near the limit of their capacity' and of driver McGrath's over-confidence in his ability to scale the summit. But most of the responsibility for the accident lay with Elliott who not only proposed dividing the train but also, in rank disregard of Company rules, had instructed the guard Thomas Henry to leave his van and to help with scotching the carriage wheels before he was satisfied that his brakes would hold. Of greater significance for future railway safety was Scrope's contention, first, that had continuous automatic brakes been fitted in the excursion train, the collision would not have happened and, second, that if a block telegraph system had been in force, its catastrophic outcome 'would have been considerably mitigated'.

In fact, as Tom would note, even before the Armagh accident occurred, there had been much progress towards the adoption of both protective devices. Nevertheless, the 1889 Regulation of Railways Act which gave the Board of Trade specific powers to compel any defaulting railway company to bring in these crucial safeguards without delay was timely. In Tom's view, the Armagh disaster was not only a legislative landmark, but also as a defining tragedy that heralded a crucial change of outlook, 'For,' he writes:

> in those shattered coaches [...] the old happy-go-lucky days [...] came to their ultimate end and the modern phase of railway working as we know it began.'[17]

Writing the book obliged Tom to reassess his somewhat less than critical view of the independent railway companies, whose idiosyncrasies he had recently and fulsomely chronicled in *Lines of Character*. One such company was the Somerset and Dorset Railway whose enterprise in securing a mainline route to the north by 'gallantly' arranging for the construction of a 26-mile extension from Evercreech Junction to Bath earns his admiring description.[18] But the project had drained Somerset and Dorset finances. Admittedly, in 1875 they entered into an arrangement with the Midland and the London & South Western Railways, but it was some time before the new regime took effect. Meanwhile, the Somerset and Dorset struggled on, its station-to-station communication haphazard; its staff few and inexperienced and its whole operation an invitation to catastrophe. Among the bleaker and more telling details to emerge from the Railway Inspectorate's enquiry into the Radstock collision of 7 August 1876, were the facts, first that the company could not afford to pay for paraffin to keep the signal lamps alight at Foxcote signal box and second, that Alfred Dando, the youth employed as signalman, did not have the strength to pull the levers. His 'training' had consisted of visiting the Radstock signal cabin while he was working as a porter to watch John Grant, the regular Radstock signalman, at work.[19]

It was August Bank Holiday and an excursion train returning from Bath to Radstock and a scheduled train heading into Bath from Bournemouth collided in the single track section of the line near Foxcote. There were at least a dozen fatalities – different accounts give different numbers – and many passengers suffered injury. In *Red for Danger*, Tom gives a very full account of the pressures, mistakes and failures of communication which precipitated the crash. Not surprisingly, at the subsequent enquiry into the accident, inadequate training of the staff was cited among its causes, but rather than attempt to apportion blame to any specific individual, the inspector Captain Tyler, reckoned that the 'whole staff were culpable [...] grossly overworked and shouldering responsibilities which they were quite unfitted by training to bear.' To their credit, the Somerset and Dorset Railway's newly constituted managing committee recognised the need for 'immediate and drastic action'.[20]

One of the book's more telling points is that the introduction of mechanical safety devices did not of itself remove the need for conscientious care on the part of the railway staff, a point which the Abermule disaster of 1921 served strikingly to illustrate. Abermule station was a crossing point between two sections of single line and to protect the single stretches, the railway relied on Tyer electric tablet instruments. The instrument held tablets which gave drivers authority to enter a particular section of track. Each instrument interlocked electrically with the one at the end of the relevant section, 'in such a way that once one tablet had been withdrawn [...] both became locked'. It was not therefore possible to release a second tablet until the first had been replaced in one or other machine.'[21] Usage of the instrument should in theory have enabled trains – whether they passed successively through the relevant section of single line working or alternated in their direction – to negotiate it in safety.

Second Breakthrough

Officially speaking, only the Abermule stationmaster and signalman were authorised to work the Tyer instruments. Had they been in the signal box, the signalman would have been able to keep them under his charge, but at Abermule, the instruments were located in the station building. Instead of their being worked by the stationmaster who had many and various duties, they were operated by any of the railway staff who happened to be on hand at the relevant time.

At 11.50 a.m. on 21 January 1921, signalman Jones accepted the stopping train from Whitchurch via Montgomery. Specifically, it meant pressing the release on the Montgomery-Abermule instrument so that his Montgomery-based colleague could withdraw the tablet giving the train's driver authority to enter the Montgomery–Abermule section of line. Next, Jones telephoned Moat Lane Junction at Caersws – some eleven or so miles' distance from Abermule – to ascertain the whereabouts of that morning's Aberystwyth – Manchester express. Informed that it was now heading towards Newtown, he returned to the signal box to set the points and signals for the stopping train and opened the level crossing gates. Although he passed acting stationmaster Frank Lewis, who was overseeing wagon movements, Jones said nothing to him about the imminent trains since at this time, there was nothing out of the ordinary to report. Not long afterwards, the Newtown signalman telephoned for permission to send the express forward to Abermule and spoke to 17-year-old porter, Ernest Rogers. Rogers, following customary Abermule practice, accepted it, and pressed the release on the Abermule-Newtown instrument to enable Newtown to withdraw the tablet. He told neither Jones nor Lewis what he had done. Francis Thompson, the 15-year-old booking-clerk, had been with him at the time, but Thompson either did not notice or did not appreciate the significance of Rogers' action.

Next, Rogers went to the ground frame to set the points for the express to enter the passing loop but found the levers locked. Before he could ask Jones to unlock them from the signal box – it was located at the opposite end of the platform - he had to attend to the passengers from the stopping train which had just arrived. Thompson meanwhile collected the Montgomery-Abermule tablet from its driver and, returning to the booking hall, met Lewis who asked where the express was. Unaware of the import of Rogers' telephone conversation with the Newtown signalman, he replied that it had just passed Moat Lane junction which led Lewis to think that it was running late. Explaining that he had to collect the tickets of the stopping train's passengers, Thompson then handed the Montgomery-Abermule tablet to Lewis. Without looking at it – he assumed that by this time Thompson would have replaced the Montgomery-Abermule tablet in the instrument – Lewis, who took it to be the tablet giving the stopping train authority to head onwards towards Newtown, handed it back to the driver. Both Thompson and Jones saw what happened, but neither raised any question.

Thompson guessed that Lewis would have changed the tablet while he had been completing his ticket round, and Jones made the assumption that for some reason the express had been delayed.

It was, in fact, approaching Abermule at speed when its driver and fireman saw steam ahead and realised that another train was in their path. Despite the driver's emergency brake application, they had no chance of preventing a collision. It killed the engine crew of the stopping train instantly, their light locomotive overrun by the substantially heavier express engine. Some fifteen passengers also died. Injured, the driver of the express wondered aloud whether Lewis had given him the correct tablet and his fireman managed to retrieve both tablets from the wreckage. The one that he found on the engine of the express was correct for the Newtown-Abermule section of the line, but the tablet carried by the stopping train engine was for Montgomery-Abermule.

In *Lines of Character* Tom had portrayed the Cambrian Railway's operation in rosy terms. 'If your train was late', he wrote:

> you did not grumble to yourself that "they" ought to do something about it. The odds were that you knew the reason for the delay as well as the staff and when the train eventually arrived, some badinage at the expense of the driver or the guard would soon restore your good humour.[22]

In the same book, he ascribed the Abermule crash to the unfathomable workings of chance; the unfortunate outcome of 'an extraordinary and most unlucky combination of minor misunderstandings'.[23] Clearly, the experience of sifting through the Railway Inspectorate's accident reports led to a radical alteration of his perspective. 'We may take it as certain', he states in *Red for Danger*:

> that the errors committed [at Abermule] ... were by no means unique. [...] At many another single line crossing station unauthorised staff had worked tablet instruments or had passed tablets from hand to hand. Many another stationmaster had ... failed to be present on the arrival of a train and then authorised the 'right away' without proper assurance that all was as it should be. Many another driver had received a tablet without examining it. Time and again, these mistakes had been made with impunity, but at Abermule ... like the scattered pieces of a jigsaw puzzle, these trifling faults fitted one into another until the sombre picture was complete.[24]

The Abermule disaster, in other words, had not come through the workings of fate, but was the result of human error. It is a rare moment in which he acknowledges that the practices of the past – specifically, the railway past – were not always benign, particularly when they involved overlooking lax procedures.

In the light of the disappointing sales of *Lines of Character*, *Railway Adventure* and, most particularly, *Winterstoke*, Tom was doubtful that *Red for Danger* would be a success. His hopes, he reflected, 'had been dashed too often'.[25] In fact, ever since its appearance in 1955, it has sold extremely well. Not only was it widely and favourably reviewed on its appearance – admittedly, a number of commentators suggested that it might not be ideal reading for a railway journey – but to Tom's satisfaction it progressed through two impressions in hardback before Pan Books brought out a paperback edition in 1960.[26] It has never been out of print and as recently as 2022 it came in for high commendation by railway operational safety specialist, Greg Morse.[27] Among the many tributes that Tom's writing has received, one of the most memorable comes from a former teacher who recalls how, facing a disruptive class of disaffected 15-year-olds, he enthralled them by reading aloud Tom's account and analysis of the 1915 Quintinshill disaster.[28] Having finished it, and with an hour-and-a half's lesson time ahead of him, to his amazement, from the class came the request, 'Go on, tell us another one, Sir'. It is an extraordinary testament to *Red for Danger*'s narrative power.

In August 1953, Tom received a letter, terse with import, from an energetic publisher named Frank Rudman, known chiefly for his championship of paperback publications. Rudman seems also to have had a certain aptitude for persuading eminent editors to accept books which had caught his liking.[29] Apparently 'Cape the Publisher' had the idea of bringing out a book about the engineer, Brunel, and had asked Rudman to suggest a likely author; Rudman's thoughts turned immediately to Tom, whom he considered 'a natural' for the task.[30] 'Cape the publisher' was in fact David Cape who had recently left the army to work for Jonathan Cape, the rather august publishing firm which his father had founded and which bore his name. While Rudman and David Cape had every confidence in the proposed Brunel book's success, *paterfamilias* Jonathan Cape was warier. Talking to him, Tom soon realised that the distinguished publisher had never heard of Brunel. His demand that Tom, whose publication record by this time was extensive, submit a synopsis and sample chapter rankled; when, eventually, old Cape came up with a contract, it proved to be 'quite the worst' that Tom had ever seen and, not unreasonably, he turned it down.[31]

It was a discouraging episode. While Tom recognised that biographies of little known figures had little chance of success, he had assumed nevertheless that Isambard Kingdom Brunel's achievements enjoyed wide-spread recognition and acclaim. Only recently, American historian James Dugan had written a book about SS *Great Eastern,* yet 'Brunel? Who's he?' was a question which Tom found himself regularly having to parry.[32] That such ignorance surrounded so compelling, so winning a figure as the designer of the Great Western Railway, the Clifton Suspension Bridge and the three great ships went entirely against Tom's grain and, despite Jonathan Cape's snub, he set about planning his Brunel biography in a surge of zeal.

Through Bristol lawyer Roy Boucher, he arranged to visit the University of Bristol library which, in the 1950s, housed Brunel's plans, sketches and notebooks for the necessary research. Boucher also put him in touch with Bristol's city archivist, the Port of Bristol Authority, and – even more significantly – with Brunel's granddaughter, Lady Celia Brunel Noble who lived in Bath. Casting his net wide, Tom also approached a Mr Lakeman, Engineer of British Rail's Western Region, who answered his queries about the Box Tunnel's rock lining; Mr Kirwan of Samuel Williams & Son Ltd, who offered to loan him copies of the *Thames Iron Works* gazette; and M. Henri Delgove, who wrote to him from Le Mans enclosing an enthusiastic piece from an unnamed French newspaper headed '*L'Extraordinaire existence des Brunnel [sic], inventeurs aux conceptions révolutionnaires*' and chattily dismissed Dugan's book as 'full of trash and gossip'. He also usefully put Tom in touch with Charles Dollfus, curator of the Paris Air Museum, an introduction which would prove immensely valuable when Tom came to write *The Aeronauts,* his history of ballooning.[33]

By this time, he had a contract for the Brunel biography with the publisher Collins whose senior editor George Hardinge, notwithstanding that his main interest appears to have been crime fiction, was highly optimistic about the book's prospects.[34] For some reason, he brought his colleague, Milton Waldman, into discussions with Tom who was unimpressed by Hardinge's lavish promises to make it a bestseller. Jibbing, Tom explained that it was not so much his intention to write a blockbuster as to produce the 'definitive biography of Brunel'. The episode, at least as Tom relates it in his autobiography, illustrates graphically his sense that money earned was at best a questionable yardstick by which to measure either the value of an exposition of engineering history, or indeed the success of any one of his books. His exchange with Waldman – himself the biographer both of Joan of Arc and Sir Walter Raleigh – ended in an impasse. Its outcome was that Tom sent the text of 'Brunel' to John Guest, now working for Longmans, who agreed not only to publish it, but also to repay Collins for their advance.[35]

In the course of Tom's research, Major Sir Humphrey Noble – Brunel's great-grandson – and his wife Celia invited him to Walwick Hall, their Northumberland home.[36] There, a number of Brunel's private papers – letter books, journals and some diaries belonging to his father – were reverentially preserved. For Tom, it was a major coup which paved the way for some rewarding investigation. What was more, the house stood on Hadrian's Wall, with moorland, forest and fell on every side and besides these delightful surroundings, his hosts were boundlessly gracious and welcoming. 'Never', he writes, 'was researcher more hospitably entertained.'[37]

Sir Humphrey Noble's own eager interest in Brunel's life enhanced the pleasure of what, to Tom, was already intensely satisfying work. While it would hardly be fair to suggest that Tom permitted the ambience of Walwick to sway his assessment of Brunel's character and career, he may have allowed Sir Humphrey Noble's impatience with John Scott Russell to shape aspects of his thinking about the management of the SS *Great Eastern*'s construction. That the two men discussed Russell is evident from a letter from Sir Humphrey which reached Tom when he was at home at Stanley Pontlarge. 'Of course Scott Russell was an extraordinary person', Sir Humphrey allows before telling the tale of Russell's acting 'as intermediary in the sale of guns, made by Sir William Armstrong at his Elswick Works, to representatives of the State of Massachusetts.'[38] If Armstrong supposed that Russell acted on his behalf, apparently one Colonel Ritchie of the US made the same assumption and 'deposited in London bonds sufficient to cover the full contract price.' Russell, the Colonel supposed, would draw them to pay Armstrong 'by instalments, less his commission, as the guns were finished'.[39] But when the full consignment of guns was complete and the bonds drawn in full, Armstrong was still owed something over £5000. Lawyers acting respectively for Armstrong on the one hand and the US Colonel Ritchie on the other 'agreed to act in concert to recover what money they could from Russell and divide the proceeds'. Russell, it emerged, had no assets on which they could claim, except for copies of his three volume textbook on naval architecture, awaiting its imminent publication date. Belatedly recognising the straits in which he stood, Russell agreed to the books' early release, but his publisher objected on the ground that, since he had not been paid for the book's production, he had a prior claim upon its author for remuneration. It was embarrassing all round, especially since Russell at this time held the rather eminent position of Vice-President of the Institute of Civil Engineers, and there ensued what Sir Humphrey terms 'a tremendous "to-do"'.[40]

If Tom took this tale at face value, it was not entirely surprising. Working at speed, he was prone to occasional carelessness over details. In *George and Robert Stephenson*, for instance, he erroneously describes canal engineer Robert Whitworth – a protégé of James Brindley – as Brindley's son-in-law.[41] But there was another factor in play. Tom, by his own admission, was much indebted to the generosity of Sir Humphrey and Lady Noble in giving him free run of their Brunel papers. At the same time, in the face of Walwick formality, he felt somewhat cowed, as his recollection of Sir Humphrey's taking him to task over the solecism of presuming to smoke while the after-dinner port was in circulation bears out.[42] To presume to question his host's story might, in the circumstances, have looked ungrateful and, instead, Tom accepted the Brunel descendants' assessment of Russell without investigation or demur. It is fair to add that although in 1867 Russell found himself obliged to resign from the ICE in the face of allegations of professional misconduct, the charges were never proven.[43]

Tom who does not seem to have got over his surprise that Brunel was not as well-known as he had supposed, wrote the biography on rather cautious lines. He opens, for instance, by showing us the young Brunel through the eyes of one Charles Macfarlane – his travelling companion in a coach bound from Paris to Calais and then on the cross-channel packet boat in 1829. Enabling us as we read to see the engineer at the start of his career through the eyes of a stranger is a deft way either to introduce him or to reawaken dormant memories of the man and his achievements. It seems reasonable also to suppose that Tom, when first Guest had aired the idea that he might care to write a life of Brunel, seized the opportunity in hero-worshipping spirit; after all, it chimed perfectly with his lifelong attachment to the GWR.[44] But adulation is not necessarily what readers look for in a biography. In time, other writers began to question the warm approbation that Tom accords to his subject. Adrian Vaughan, for instance, author of a life of Brunel which appeared in 1991, objects to what he terms the 'cosy, romantic tone' of Tom's writing. Quick to disprove Tom's assertion that no strikes accompanied the GWR's construction, Vaughan is patently uneasy with his disparaging portrayal of John Scott Russell.[45] While Brunel's recent biographers, Angus Buchanan (2002) and Steven Brindle (2005) do not share Tom's critical view of Russell, neither, for the record, do they follow Vaughan in exonerating him.

Vaughan's claim, indeed, that Tom portrays Brunel with 'uncritical hero worship' merits investigation.[46] By the mid-1950s when he wrote the book, Tom had received enough correspondence from readers of his earlier publications to have a fair sense of his audience. In the main they were committed car, rail or canal enthusiasts – quite possibly VSCC, TRPS or IWA members, who viewed him, if not as a friend, then certainly as a like-minded acquaintance whom they could trust to write engagingly on engineering subjects. While such readers might enjoy some occasional sense of surprise – the tale of Brunel and the swallowed sovereign, for instance, is nothing if not piquant – they were not looking for shocking revelations. In his classic *Lives of the Engineers*, nineteenth century writer Samuel Smiles portrayed his subjects in a broadly heroic light and as exemplars of the merits of self-reliance.[47] Tom rather liked the idea that he had inherited Smiles' mantel and in an article for the journal of the Society of Authors, he presented himself as following in Smiles' footsteps.[48] He brought more solid engineering knowledge to his books than Smiles could claim to possess, which is evident in his discussions of the technical aspects of designing bridges, for instance, or the principle on which the Atmospheric

Railway was intended to run. At the same time, like his Victorian predecessor, he palpably enjoys the business of telling a good story and if he makes much capital out of Brunel's regular trouncing of his enemy Dionysius Lardner, it adds to the general narrative zest.

That it chimed both with the spirit of the time and with readers' expectations is evident from the success not only of the Brunel biography, but also of its successor *Thomas Telford* (1959) which earned the special recommendation of the Book Society and *George and Robert Stephenson* (1960) which completed the trilogy. Noting that the Book Society did not bestow the imprimatur of its recommendation upon *Brunel* led Tom to quibble about being a 'victim of Sir Charles Snow's "two cultures"' which could not regard an engineer's biography as a literary production, but time would show that his fretting was misplaced.[49] In 1965, Newcastle University recognised his work with the award of an honorary MA degree.[50]

That all three titles should have had the effect of promoting new research into the respective engineers' lives speaks eloquently of the appetite for further information that they inspired. Not only have Vaughan, Buchanan and Brindle produced stimulating books about Brunel in the wake of Tom's 1955 publication, but the early 1970s witnessed the repatriation of SS *Great Britain* to her home port of Bristol after her abandonment in the Falkland Islands. Restored, the ship is now the nucleus of a flourishing museum and Institute dedicated to research into her creator's life and achievements. The Stephensons, always popular with writers and readers, caught the attention of journalist Simon Garfield – winner in 1994 of a Somerset Maugham award – whose account of the inception and opening of the Liverpool and Manchester Railway, *The Last Journey of William Huskisson* appeared in 2002.[51] As for Thomas Telford, whom Tom purported – rather harshly – to find 'dour, reticent and solitary', he is the subject of a resplendent biography by Julian Glover which came out in 2017 and was a BBC Radio Four Book of the Week.[52] That engineering lives should feature so prominently in twenty-first century publishing would probably have surprised Tom and assuredly it would have delighted him. It is a phenomenon for which he was largely – if indirectly – responsible.

Chapter Fourteen

Chronicler of Companies

Besides his studies of Brunel, Telford and the Stephensons, Tom cast his biographical net wide. Seeking to argue that if in truth the locomotive had an inventor, the Cornishman Richard Trevithick had a better claim to the title than George Stephenson, before writing his Stephenson biography, he wrote a concise if searching account of Trevithick's life. Lutterworth Press – not a company with which he had many dealings – published it in 1960 supposedly as a children's book with the engaging if less than explicit title of *The Cornish Giant*.[1] Two years later, Tom's *Great Engineers* appeared – a collection of concise life stories of ten innovators, ironmaster Abraham Darby being the earliest and motor car and aeronautics entrepreneur Frederick Lanchester the most recent. The others, chronologically, were steam engine pioneer Thomas Newcomen; canal man William Jessop; mechanical engineer Matthew Murray; precision tool specialist Henry Maudslay; railway builder Joseph Locke; early originator of the traction engine John Fowler; civil engineer and bridge designer Benjamin Baker and Rookes Crompton, designer of dynamos. What linked them, Tom argued in a spirited Foreword, was the fact that although their achievements were 'just as significant, historically, as those of their more celebrated contemporaries', they had somehow slipped out of the public gaze. The experience of seeing 'the same few famous names [...] over and over again' on the biography shelves even in specialist libraries was, he complained, 'depressing' and so he had set out to remedy the deficiency.[2]

In some circles, the resulting book found considerable favour. Gilbert Thomas, the *Birmingham Post* reviewer, applauded its insights and the *Daily Herald* featured it in its 'Book A Day' column for 9 February 1962, which opened with the reviewer – he happened to be the Principal of the Manchester College of Science and Technology – announcing that it was 'wonderful to be reminded that English [sic] engineering was once the best in the world'.[3] What impact it made with the public at large is hard to say, but the publishers, George Bell & Sons, evidently thought it worth their while to issue a reprint four years after its initial appearance.

His final engineering life was the biography of James Watt which appeared in 1962. Batsford who had dithered years previously over *Narrow Boat* and more recently commissioned his book about the Thames published it in their 'Makers of Britain' series. Compared with his Brunel, Telford and Stephenson books, it was rather a lacklustre production. Tom's irreverent contention that 'compared with [George] Stephenson, Watt is an also-ran in popular memory' holds some shrewd biographical insight. The kettle anecdote may have lent lustre to Watt's name and reputation, but his work on the steam engine never, perhaps, engaged people's imaginations in the way of either George Stephenson's locomotives or Brunel's bridges and great ships. While he traces Watt's life from Scotland to London, Birmingham and Cornwall, Tom's writing gives a sense of disengagement from his subject, as though he is never entirely at ease in Watt's company. An unusually gleeful letter that Watt wrote to Matthew Boulton in June 1779, having concluded a tedious argument with Hawkesbury Colliery, leaves Tom not only unmoved, but positively embarrassed. From Watt's exclamations – 'Hallelujah! Hallelujee!' – his biographer recoils discomfited, as though he were witnessing 'an elder of the Wee free dancing a hornpipe at a funeral'.[4] Never widely reviewed, *James Watt* does not appear to have been reprinted since its first publication.

Early in the 1950s, the publishers Newman Neame Ltd invited Tom to join the 'panel of authors' on whom they called to write company histories in limited editions for private publication.[5] His earliest work in this area was a self-contained historical chapter in a book which the lightering company Samuel Williams & Sons Ltd commissioned to mark their centenary. The work chimed well with his interests. Samuel Williams himself, who established the company in 1855, had a mix of opportunism and resourcefulness which Tom admired. Initially, Williams had had no onshore base, but made his deals in the London pubs and coffee houses and operated his lightering vessels from his Thames moorings. In time, he took out a lease on Princes Wharf in Battersea and, having enough business sense to recognise an opportunity in the making, took the chance of buying land on the Dagenham Level in the 1880s.

A marshy expanse within protecting river-walls which had broken from time to time over the years to admit a flood of water from the lower Thames, it was an area where engineering history loomed large. From his reading of Samuel Smiles, Tom knew about the work of Cornelius Vermuyden who had stemmed a significant breach in one of the Thames embankments in 1621 and Captain John Perry who closed a new breach with wood piling when it opened in a 1707 storm.[6] For over a century low-lying Dagenham Level had remained undisturbed behind the bank which protected it from the Thames but, when railways expanded east out of London, its potential as the site of a useful dock became increasingly difficult to ignore. A dock company had formed in the 1860s, engaged contractors to develop the old gulf and recruited Sir John Rennie to direct the works. Lacking capital the venture failed, but Williams had been quick to see the area's potential and purchased the site himself in 1887.[7]

To make the best of its advantages, he had to build his centre of operation pretty much from the beginning upon the expanse of marsh. The development of the harbour for handling rubble, or rather the 'vast tonnage of subsoil' that the excavation for the London Underground system produced, was the first step in a process of land-reclamation for the purpose of transforming the mudflats into a harbour. By the 1890s, Williams' operation had conclusively moved to Dagenham while a brewery moved onto the old Prince's Wharf site. 'A martinet' in Tom's opinion, Williams 'ruled both his family and his business with a rod of iron'. Nevertheless, when Tom visited the Dagenham Dock, the centre of Williams' operations, the company's friendliness impressed him and he could not help but notice the existence of a 'vital personal link between management and men' which, he observed, 'can [...] easily be lost when a business grows'.[8]

The connection with Samuel Williams & Sons enabled Tom to fulfil his cherished ambition of making a long sea voyage in a working ship. A chance cocktail party conversation with John Carmichael, Samuel Williams' director of the time, resulted in an invitation to the Rolts to travel out to Barahona in the Dominican Republic aboard the company's new bulk carrier *Hudson Point*. Surprising as it may seem for a writer who so often draws upon his memories, Tom was not a regular diarist in the sense of James Boswell, for instance, or Francis Kilvert. For this voyage, however, using the pocket-sized Timothy White 1958 appointment diary in which Sonia and he noted their various engagements, he recorded his day-to-day recollections of life at sea and later drew on them to furnish the account of the adventure which appears in *Landscape with Figures*.

Having left their two small sons in the care of a nanny, Tom and Sonia joined *Hudson Point* in John Readhead & Sons' South Shields shipyard on 27 January 1958. Since the vessel did not 'officially' take passengers, to comply with the regulations they agreed to sign on as steward and stewardess. There was, apparently, no expectation that they would undertake any of the duties associated with the roles.

Hudson Point was one of two ocean-going vessels that Samuel Williams' Newcastle-upon-Tyne subsidiary, the Hudson Shipping Company, acquired in the mid-1950s with a view to engagement in the West Indian raw sugar trade.[9] Having cast off late in the afternoon of 29 January, by 11.30 p.m. on the following day, they had rounded North Foreland and passed Dungeness. On the morning of 31 January, in mist they passed the island of Les Casquets with its lighthouse, and found themselves off Ushant by the afternoon. As they headed west, the calm of the early stages of their journey gave way to something rougher; 'gentle swell' wrote Tom on 1 February, adding the details, 'saw porpoises in the afternoon [...] Automatic pilot went wrong'.

The weather worsened, leading Sonia to take to her berth; 'so long as she remained in a horizontal position she felt alright', Tom recalled. He, meanwhile, remained upright despite the constant rolling, pitching and pounding, although the sheer exertion involved in keeping his balance left him exhausted. In calm conditions, he and Sonia would sit out on the bridge deck in deck chairs, much like passengers enjoying a cruise, but in the early stages of their voyage, such moments of respite tended to be brief. At the end of the first week of February, came a day of heavier than usual swell 'with squalls and thunder' and force 6/7 winds; 'sometimes almost lea rail under', Tom ominously observes. The night was no better and his scrawled diary note about its being 'most uncomfortable' may have been an understatement.

But the next day was bright and sunny and although the ship continued to roll, Sonia recovered her sea legs. On 9 February, the Captain gave them a thorough tour of the ship and after tea, Tom took the chance to visit the engine room. By now, the sea was calm and they saw their first flying fish. 11 February, a day of hot and sunny weather, was Tom's birthday which Sonia and he marked with a bottle of Burgundy over dinner. A 'small German ship' and a Clan Line vessel passed them, and they also noticed a salvage tug – the first ships that they had seen since leaving the English Channel. Looking astern, they saw phosphorescence gleaming in *Hudson Point*'s wake. The next day, a couple of US Destroyers passed them in the morning and around mid-afternoon, they sighted land. Around noon on Thursday 13 February, observing the port of Barahona through his binoculars, Tom saw two previously recumbent figures embark upon a small white launch while a siren wailed and realised that the local pilot was on his way. Once in the harbour, the ship's crew at once set about loading their cargo of raw sugar.[10]

It was a lengthy task. At the time, Barahona had no equipment for handling bulk goods, but relied solely upon a meter-gauge Baldwin wood-burning steam locomotive with a balloon chimney which, Tom reports, 'shuttled to and fro between the quay and [the] sugar mill freighted with stacks of bagged sugar'.[11] The kindly Spanish manager of the Sugar Mill arranged not only for Tom and Sonia to visit the plantations but also introduced them to Mrs Grant, widow of a former sugar plantation owner and Barahona's sole English resident. They also visited what is now San Domingo, known in the 1950s as Cuidid Trujillo. Having rebuilt the city after its destruction by the 1930 San Xenon hurricane, dictator Rafael Trujillo decided to rename it after himself.[12] But Tom, whose whole purpose in making the journey derived from his wish to 'travel the ocean the hard way by a merchantman' while taking the chance to explore his ship all over 'from bridge to engine-room' was not in tourist mood. Sonia and he entered into their shore excursions with good grace, but he admitted outright that, although he thought the Dominican Republic trip a 'worthwhile experience', it was not one which he would care to repeat.[13]

By the early 1960s, the business of compiling company histories served Tom as a regular and financially useful sideline to his more mainstream writing. Typically, a firm would commission the book to mark a significant anniversary of its existence signalled in a subtitle such as 'the first fifty years', or 'growth over a century'. Often, the work that it entailed held its own intrinsic interest. In *Landscape with Figures,* he describes a visit that he made to Whitehaven in the course of writing *Mariners' Market* to mark the centenary of ships' store dealers Burnyeat Ltd. Although the company later established its headquarters in Liverpool, Whitehaven had been where William Burnyeat made his initial foray into commerce and where, as Tom discovered, he had opened a modest chandlery business in December 1851 – a full decade earlier than anyone connected with the firm realised, since Burnyeat Ltd blazoned its origins proudly if inaccurately on letter-headings and trade vans bearing the palpable legend 'Established 1861'.[14] Quite what exchanges Tom, who guessed that he owed his commission to the notion that 1961 marked the firm's centenary, may have had with the Burnyeat directors at this point is not, sadly, something that he cares to disclose.

Nevertheless, he enjoyed the time that he spent in Whitehaven and in the course of some enterprising field-research, he discovered William Burnyeat's original business premises behind a drapers' shop on King Street. He also found the purpose-built on-shore accommodation that Burnyeat provided for his 'captains, cronies and customers' while their ships were in port.[15] Another unanticipated revelation came when, peering inquisitively through the cobwebby window of a hut on one of the quays, Tom saw the carved figurehead of company ship *Emily Burnyeat*, built by Gowan of Berwick in 1862 and persuaded its owner to allow him to photograph it for his book's cover.[16] To Tom, a little knowledge of a subject always served to spur his zeal to learn more about it and the adventures that he had in the course of compiling *Mariners' Market* came about through the imagination and zest that he brought to what, in the hands of a less engaged writer-researcher, might have been rather an arid task.

Since Newman Neame's books often appeared without any mention of the author's name, it is not easy to be certain quite how many company histories Tom actually wrote. Besides his contribution to *A Company's Story in its Setting: Samuel Williams & Sons Ltd 1855-1995* and *Mariners' Market*, Peter Roberts, in his recent Rolt bibliography, credits him with both *Holloways of Millbank – the First Seventy-five Years* and *The Dowty Story*.[17] This list may not be exhaustive. Tom mentions, for instance, writing the history of a London-based civil engineering contracting firm whose sometime civil engineer-in-chief had overseen the 1935 dismantling of the old Chelsea suspension bridge. Tom apparently paid the engineer an eventful call during which, to his surprise, he found himself assisting in the delivery of an Aberdeen Angus calf. Animal-husbandry aside, in the course of Tom's visit the engineer related how, during his work on Chelsea Bridge, he had come up with the idea of casting cables across the Thames to support timber staging and pulling them taut so that the staging took the weight of the old bridge's suspension chains, enabling the removal of their link pins. At the time, his objective had been to take the chains apart without unduly disrupting the river traffic. Tom's response to this tale of on-the-job initiative was, apparently, to remark that the same ingenious idea had suggested itself more than a century earlier to Thomas Telford at Conway. If the comment left the Chelsea Bridge engineer 'somewhat dashed', it is perhaps not entirely surprising.[18]

Newman Neame was not the only publisher with an interest in bringing out this type of book. Tom's histories of agricultural engineering company Taskers of Andover and locomotive manufacturer Hunslet of Leeds evidently held enough general interest to be taken up by new publishing company David & Charles, which Tom's friend Charles Hadfield had established with railway enthusiast, David St John Thomas.[19] Having come into being on 1 April 1960, the firm swiftly cornered the increasing market for railway and canal-related books and other titles of historical engineering interest. When, at Tom's instigation, they reissued Temple Thurston's 1911 canal classic *The Flower of Gloster*, he provided the new edition with its introduction.[20]

While he clearly enjoyed digesting company archives, meeting long-serving staff and recording the annals of different firms' commercial experience, inevitably the relationship between the organisation on the one hand and its chronicler on the other did not always work out. Tom had relatively little managerial experience and what he possessed, he had largely gained within the idiosyncratic context of the Talyllyn Railway, but he was not lacking in worldly shrewdness. 'A commercial enterprise', he observed:

> has its ups and downs, its faults and its failures as well as its successes, and any history that pretends otherwise must not only ring false, but also makes very dull reading.

Blips or downturns in a company's fortunes were, in his view, integral to its history and grist to the chronicler's mill and he was not prepared to countenance any request from the firm's management to conceal them. The directors 'of one large group' of companies apparently raised objections to his projected history-in-progress of the consortium because it included an account of a destructive workplace explosion. That the blast had occurred in the First World War, long before the group absorbed the company concerned, evidently counted for nothing with the Board who proposed that Tom omit all mention of it and cut his history by 10,000 words. Since his initial contract had specified that his text was to be in the region of 60,000 words – a modest length, given the organisation's size and scope – Tom decided, at this point, to bow out, with the suggestion that project might best be entrusted to the company's public relations department. To their credit, despite this rebuff, the company paid his fee in full.[21]

Tom does not mention it in *Landscape with Figures*, but a commission to write a history of asbestos manufacturer Turner and Newall Ltd confronted him with a different species of dilemma. By the 1950s, if not earlier, the dangers that asbestos presented were becoming apparent. Cherishing their reputation as a sound employer and keen to avoid any adverse publicity, Turner and Newall had grown rather deft in discrediting the growing body of evidence concerning asbestos-related diseases, notably mesothelioma. But in 1964, the firm's solicitors advised the directors that they would not be able to talk their way out of asbestos-related insurance claims for ever. The epidemiological knowledge of the risk that working with asbestos presented was growing too fast.[22] Tom completed his history of the company and by 1970, *Turner and Newall Limited: the First Fifty Years, 1920–1970* was printed and to all appearance ready for circulation, but it was never officially released. Whether the decision owed more to Tom's conscience or the company's commercial scruples and reluctance to make themselves a hostage to fortune, it is impossible to say. Tim Rolt – Tom's son – and Rolt bibliographer, Peter Roberts, suggest that the probable reason behind the book's suppression was the 'adverse publicity' which, by the late 1960s, had come to surround the industrial production of asbestos and its use in building.[23] Nevertheless, copies of the unpublished work find their way onto the market from time to time.[24]

Despite the success of his bestselling titles, notably *Narrow Boat*, *Red for Danger* and the biographies of Brunel and the Stephensons, Tom remained acutely conscious that living by his pen was always something of a gamble. For an income, he relied on royalty payments and the fees that he received for the company histories, together with intermittent earnings from lecturing, book reviewing, journalism and the occasional BBC broadcast. In a splenetic passage from the 'Truth about Authorship' chapter of *Landscape with Figures*, he inveighs against numerous factors which resulted, in his view, in authors' receiving less than fair remuneration for their work. Among them, he lists publishers' 'anachronistic' accounting methods; quirks of the Inland Revenue; booksellers' reluctance to stock or pursue non-mainstream titles; pressure from publishers to accept ever lower percentage royalties and – touching upon a situation which held sway in the early 1970s, but which no long exists – the absence of any public lending right payment scheme.[25] That he, as he boasts, 'succeeded in keeping a wife, bringing up and privately educating two sons, assisting an impoverished mother and maintaining an ancient Gloucestershire house' from his literary earnings was, in his view, a reason for much pride.[26] In the twenty-first century, to maintain such a lifestyle on a writer's average income would be impossible. That Tom managed it in the 1960s and 1970s, owed much both to his tenacity – not to say bloody-mindedness – and to the quirk of fortune which ensured that a market existed for the type of books that he enjoyed writing.

He was prolific, by any reckoning. Concurrently with the compiling company and commercial histories he also produced *Thomas Newcomen: The Prehistory of Steam* (1963), an account of the early history of the steam engine; *Patrick Stirling's Locomotives* (1964), a study of the work of Patrick Stirling, sometime locomotive superintendent of the Glasgow and South Western Railway; *The Aeronauts* (1966), a history of ballooning; *Navigable Waterways* (1969) for the Longman Industrial Archaeology Series, of which he was the General Editor; and *Victorian Engineering* (1970), an overview of nineteenth-century technological development. Considering that at the same time, he gave lectures, reviewed books and contributed newspaper and journal articles, it made for a prodigious workload.

Ever since the campaigning canal cruises that he had undertaken in the 1940s and 50s, Tom had been concerned by the condition of the early industrial buildings that he encountered on his travels. While he conceded that the untrammelled commercial energy which had brought into being mines, mills, locks and bridges had its evils – 'disastrous ... social conditions; and desperate overworking and under-payment of the labour force' among them – he stood in awe of the 'superb and painstaking craftsmanship' of their construction.[27] He was not alone in his thinking. At the end of the 1950s, the Council of British Archaeology convened a conference dedicated to the emerging discipline of industrial archaeology which Tom attended. It brought together many people who shared not only the same broad interests but also the same concerns. Specifically, by the end of the Second World War, Britain's transport and industrial infrastructure had not only sustained much damage but, it soon became apparent, it was no longer adequate for its purposes. Town planners began to harbour ideas of 'sweeping away [...] the old order' to replace it with something that was visibly innovative – something that spoke of modernity. A phase of 'rapid destruction of old industrial structures and urban environments' ensued, which reached its apogee with the 1962 demolition of the Euston Arch – the Doric Portico entrance to Euston Station, which was at the time undergoing a thorough renewal and redesign.[28] Despite a public outcry and some energetic campaigning by poet and vice president of the Victorian Society John Betjeman, with whom Tom had been on friendly terms ever since their 1950 meeting at Buscot Rectory and who spearheaded a vigorous campaign to relocate the Arch at a cost of £90,000 to be raised by public appeal, Minister for Transport Ernest Marples declined to sanction its preservation.[29]

Collectively perceiving that the drastic renovation in progress was destroying many buildings of value as well as the architectural dross, interested members of the public took stock. In 1964 Angus Buchanan, at the time an assistant lecturer at the Bristol College of Science and Technology, perceived that 'many metaphorical babies were being thrown out with the bath water' and set up a satellite Centre for the Study of the History of Technology at Bath University appointing Tom to its advisory Council.[30] The Centre was quick to survey industrial monuments, initially focussing on those in the Bristol area, and to publish a *Journal of Industrial Archaeology* – in other words, it set industrial archaeologists' work onto some rigorous footing. Not only would it keep records of their finds, but it began to organise regular conferences – a delight to Tom who always enjoyed meeting and talking to people who shared his interests. Bath University recognised his contribution to the expanding discipline of Industrial Archaeology with the award of an honorary MSc degree in 1973.[31]

How far he relished the administrative business of the Centre's Council, or indeed, any committee work, is more questionable. In time, among other committees, he would serve on both the Inland Waterways Redevelopment Committee and the Science Museum Advisory Council. While in *Landscape with Figures*, he allows that such bodies 'do a good job', it sounds as though he may have viewed his committee work with greater sense of duty than zest. 'As we grow older', he writes, 'We tend to talk more and to do less and, for readers of biography or autobiography, the doing is always more interesting than the talking'.[32] It is a fair point, although his committee work would be pivotal in bringing the present National Railway Museum into being. Sadly, it did not open until after his death.

The late 1960s found him appointed Editor in Chief of Longman's Industrial Archaeology Series which included works on such diverse subjects as roads and vehicles, the chemical industry and building materials.[33] It also included his own *Navigable Waterways* (1969) and in his introduction, he quoted the definition of civil engineering which Thomas Tredgold supplied in 1828, namely that it was 'the art of directing the great sources of power in nature for the use and benefit of mankind'. These were resonant words, but the differences between eighteenth-century canal engineering and the construction of a 1960s motorway led Tom – notwithstanding that he wrote booklets on the construction of the M1 and the Tyne and second Mersey Tunnels – to draw some discomfiting conclusions about the direction that his profession was taking.[34] Specifically, a pressing question confronted him about whether, over the course of a couple of centuries or so, humanity's attitude towards nature had not radically altered. 'Whereas once [man] was concerned with art to direct', he argued:

> now he presumes with strength to command. What began as a sensitive and loving partnership has ended in a brutal and arrogant rape.[35]

The harsh, disturbing imagery is, to his thinking, all that fits an age in which the 'argument of expediency' – doing what is convenient rather than what is right – admits no challenge.[36]

By the time that he came to write the closing chapter of *Victorian Engineering* (1970) his thinking had grown even more desolate. Here he returns to the frustration which had been with him ever since his disastrous spell of employment with Rolls Royce performing tasks from which all acquired skill had been stripped with the result that operatives were paid for the 'monotonous repetition of a few simple operations'.[37] To an educated workforce, he reasons, such tedious employment will be unsatisfying no matter how generous the remuneration. The likelihood of some future 'collision' between the operatives and 'the system' was, in his view, a looming threat.

Since he always possessed broad interests, it is no surprise that Tom should have held up the way that members of the eighteenth century Lunar Society might, in the course of speculative conversation, 'range freely and delightedly over the whole field of human knowledge' as an ideal exercise of human intellect. By contrast, he deplored what he saw as the growing tendency to extend the principle of 'division of labour' – that is, the practice of splitting a large task into small elements, each assigned to a separate labourer – into the intellectual sphere. It served, in his view as a recipe for learning more and more about less and less.[38] Crucially, he envisaged the result of this principle of sub-division and separation of tasks as producing as 'a terrifying failure of the sense of individual responsibility'; a sense that humanity no longer had any control over, or even purchase, upon its own destiny. Instead, the process of manufacture which man had developed assumed a life of its own; not only did it follow its own path, but it also set its own agenda. Tom had seen the pattern and outcome of its workings years earlier when, as a young man, he wrote 'Strange Vista.'

'I honestly believe the whole order of things as we know it will crash', confides his *alter ego* David Hancock to Anne Cray as they walk up on the hillside above Wolvercroft. 'It's hard to explain,' he admits:

> but I am sure of it and as I have never believed in the industrial system, I don't intend to stand by it like the captain of a sinking

ship. I want to go with you and leave it all behind. Leave all the filth and misery and blackness and get away from the wheels of this awful machine before it crushes us. We can't avert it and if we stay, it's bound to destroy itself and us with it.[39]

As a conclusion to a discussion of the engineering of the Victorian era – the period when it often seems that its skill and flair reached their zenith – this vision of humanity's being in effect dehumanised by enslavement to its own creation is desolate indeed.

Chapter Fifteen

The Realm of Ghosts

Tim Rolt remembers his father's writing routine well. He writes, 'I think he was very self-disciplined. He worked from home and I have strong childhood memories of the sound of the clack-clacking of the typewriter. I think he composed directly on the typewriter. There was a lot of smoking involved....' Tom recruited Tim to be his messenger, paying him 6d for posting letters, and 1d for calling him down for a meal.[1]

While the recollections are fond, Tom knew that his choice of career made for a 'precarious' existence, and that the need to write prolifically and fast was paramount. In a 1954 letter to Hadfield, he remarked on his good fortune in being asked to organise the Anglo-American car rally which apparently 'brought in welcome shekels'.[2] Since royalties were his chief source of income, he was quick to recognise that building up a substantial back-list of books which promised to sell steadily over the years held greater advantage than production of a 'runaway bestseller' which, he observed, could be 'disastrous' to its author 'from a tax point of view'.[3]

Aside from royalties on his books, journalism was also a potential source of income. Among his most successful newspaper contributions is the delightful series on 'Midlands Monuments' – concise accounts of historic industrial structures of the Midlands area – which appeared in 1968, complete with illustrations in the *Midland Magazine*, a weekend supplement to the *Birmingham Post*. Book reviewing and lecturing he found less satisfying, regarding both activities as time-consuming distractions from writing his current book-in-progress. A contracted book, after all, even if it achieved only modest success on its first appearance, might well come into its own later. A lecture, by contrast, despite the preparation that it necessitated and the time-consuming travel that it was likely to involve, was, in his view, essentially 'ephemeral'. So, in his view, were the book reviews that he wrote. 'The most I can say for reviewing is that it is a way of building up a modern library on one's subject', he witheringly remarked. Neither lecturing nor reviewing made any significant contribution to Tom's earnings, and he describes the payment – assuming that it was offered – variously as 'paltry' or 'insultingly low'.[4]

Despite these grumbles, by the early 1970s publishers besieged him with ideas for new books and his standing as an author meant that he was free to exercise considerable freedom over choosing which ventures he cared to pursue. Sometimes, it resulted in a sort of literary cat-and-mouse game. In June 1973, Susan Kennedy from Faber and Faber's editorial departments wrote to him with the idea of a possible book about early ironmasters which she hoped that he would write. In response, Tom hedged and, having drawn her attention to Alan Birch's 1967 *Economic History of the British Iron and Steel Industry 1784–1879*, expressed doubts about whether he was the best person to write it. He also explained that he was busy, but promised to let her know if he could think of any other writer for the task. Far from taking this response as the polite negative that Tom may have intended, Mrs Kennedy thanked him for his 'encouraging remarks' about the venture, reiterated her view that it would be 'marvellous if he were to write the book himself' and even asked him to propose a delivery date for the manuscript, promising that Faber were ready to wait three years for it, if need be. Ten days later, she sent a follow-up letter suggesting that he might care to discuss that book over lunch with her when next he was in London. Despite the pushy overture, Tom was reluctant to dismiss the project out of hand. Instead, he seems to have stalled because Mrs Kennedy, who was nothing if not tenacious, contacted him again in November 1973, hoping that he had given further thought to the 'Ironmasters book', and particularly to the idea of writing it himself. What Tom would have made of the work, we shall never know because within a few months Susan Kennedy left Faber and Faber and plans for the book fizzled out.[5]

At much the same time, Charles Hadfield mooted the idea that Tom and he might collaborate in writing a double biography of engineers William Jessop – he was one of Hadfield's heroes – and his close contemporary, John Smeaton.[6] Not surprisingly, he proposed that David & Charles should bring out the book but Tom had reservations about negotiating with Hadfield's co-publisher David St John Thomas. 'I don't see why, apart from the contract, you should deal with DT at all', observed Hadfield at the time, but as things turned out, although David & Charles made an offer for the book, their terms were so unsatisfactory that even Hadfield concluded that Tom and he would do better to look elsewhere.[7] As it happened, in December 1970, Tom had secured a contract for a dual biography of Smeaton and Jessop with the Longman Group. The proposed book never materialised, but Judith Niechcial, daughter of civil engineer Sir Alec Skempton, recalls that Tom eventually gave Hadfield his notes on Jessop, such as they were, and the contract was cancelled.[8] In the event, David & Charles published *William Jessop* in 1979 – a collaboration between Hadfield and Skempton.[9]

Publishers Ward Lock had the idea of reviving the 'Wonder Book' series which had fallen out of production and David Platt, one of their editorial team, called Tom to discuss the possibility of bringing out a new railway edition. Since Tom, as Sonia observed, was 'not really a telephone person,' Platt travelled down from London to Evesham where Tom met him and took him on to Stanley Pontlarge. In the course of what he considered 'a very pleasant day,' Tom and he concluded that there was certainly scope in the market for a new *Wonder Book of Railways*. Soon afterwards, Tom sounded out his old Talyllyn friend John Snell – who by this time was Managing Director of the Romney Hythe and Dymchurch Railway – about the proposal. Snell was agog with enthusiasm and quick to pass comment on Ward Lock's provisional synopsis, quibbling over their subject headings and disputing over which topics did or did not merit a 'special chapter' of the finished book. In all likelihood, except for the fact that it took place on paper rather than face to face, it was just the kind of conversation that the two men had enjoyed many times in the past. Sad to say, the new *Wonder Book of Railways* – a project whose initial discussion phase went ahead in such zestful spirits – would never achieve publication.[10]

But not all of Tom's ventures of the time would remain unrealised or unfinished. By 1972, Heinemann Educational Books Ltd had recognised the growth of interest in Industrial Archaeology and invited Tom to contribute a chapter on inland waterways to Brian Bracegirdle's prestigious *Archaeology of the Industrial Revolution*. Not only was the commission flattering – it is difficult to think of any other writer of the time who was quite so well placed to carry it out – but it carried a generous fee of £100.[11] Some months later, Ruth Thomson of Macdonald Educational requested Tom's advice – for which her firm undertook to pay – on preparing a book intended for 6-8-year-olds about George and Robert Stephenson. She telephoned him in the autumn of 1973, but either Richard or Tim took the call. A brief but impressively comprehensive note shows just how thoroughly the boys understood the demands of their father's career, confusion around the book's subject notwithstanding. 'Dad' it runs: 'Ruth Thomson wants you to ring her up about some children's version of one of your engineer biographies, Brunel I think, that she wants to write'. A London telephone number complete with extension follows.[12]

Having recently revisited Britain's canal network for the purpose of writing *Navigable Waterways*, in 1971, Tom turned his attention to the Canal du Midi. This triumph of French engineering preceded James Brindley's work of constructing a canal for Francis Egerton, Third Duke of Bridgewater by some ninety years or so. Having conceived his scheme to connect Sète on the Mediterranean with the Garonne at Toulouse to access the Atlantic port of Bordeaux without the need for a long sea-passage, engineer and tax collector, Pierre Paul Riquet, gained the support of Louis XIV for his waterway in the mid-1660s. The fact that the third Duke of Bridgewater visited this French canal as a young man gives it at least some connection with the British canal network.[13] Although he describes himself in the book's introduction as a 'stay-at-home Anglophile', for Tom to view it with curiosity was not surprising. On 2 April 1971, together with Sonia, Richard and Tim, he took the overnight ferry from Southampton to Le Havre whence they travelled on south to Sète. Having boarded their hire boat at Castelnaudary on 6 April, over the following week they travelled to Marseillan. Admittedly, the canal closed to traffic on the Sunday, which was Easter, but it was nevertheless an opportunity to gain first-hand impressions of its construction and character. Tom was especially appreciative of the chance for some close-quarters examination of the system of feeders which supplied water to its summit level.[14] The following September, the guest of Conservative politician Robert Grant-Ferris, he returned for a more expansive westward voyage, and cruised aboard Sir Robert's motor yacht *Melita* from the port of Agde to Castets-en-Dorthe.[15]

The experience of seeing the ongoing works intended to enable the canal to accommodate vessels of 350 tonnes capacity left him with mixed feelings. On the one hand, he supported the idea of canals performing the very function for which they were constructed, namely the transport of bulk goods. At the same time, having experienced a power failure when passing through the locks between Castets and Agen which had meant working them by hand – it proved a 'laborious […] operation'— he questioned the wisdom of relying on electricity for keeping the enlarged locks adequately supplied with water. Predictably, the argument behind the new scheme turned upon economics. 'The cost of electric power in France is so low', Tom explains:

> that current engineering opinion holds that the cost of pumping back lockage water electrically is less than that of providing and maintaining elaborate gravity feeding systems.[16]

It is a not uncharacteristic reflection in which Tom conveys his admiration for the work of Riquet and his assistant Francois Andreossy, while at the same time distancing himself from the views of the French engineers of his own day.

Writing with his usual speed, he completed *From Sea to Sea* in 1972 and corrected the book's page proofs and index in the same year.[17] But the patterns of his life were about to change. A brief entry for 11 April 1972 in his engagement diary reads, 'Ill evening' and the entries for 15 April – 'to hospital, evening' – and 20 April -'Operation' – tell their own tale.[18] The nature of Tom's illness, however, and the reason for what appears to have been emergency surgery is not specified, but its seriousness necessitated cancellation of a lecture on Britain's metal bridges that he had been due to give to members of the British Association for the Advancement of Science at the end of the month.[19]

From this time on, the name of 'Fairgrieve' begins to appear in Tom's engagement diary.[20] John Fairgrieve (1926-2013) was a vascular surgeon who lived and worked in Cheltenham; that Tom had a number of appointments with him between May and December 1972 suggests that his trouble appeared to be some form of vascular disease. He would have a second operation on 7 September, and cancelled all the lectures that he was scheduled to give over the autumn.[21] Dick Harper, a GP with whom Sonia and he had built up a friendship founded on their mutual fondness for Llanthony and the nearby Black Mountains, wrote to Tom at the end of the year expressing sound medical approval for his decision to give up alcohol.[22] Tom, although often opinionated in his writing, was always reluctant to mention his health but Dr Harper's letters reveal that by this time he had also developed diabetes. In *Landscape with Figures*, he relates that friends from the past rallied round and a number of his VSCC acquaintances arranged and financed a three week holiday for Sonia and himself in the Algarve, well away from the bleak Gloucestershire winter.[23]

Even before his illness showed itself in its full seriousness, Tom had decided to write about his life. Approaching the end of his life, he observed self-deprecatingly first that the task was 'an act of self-indulgence' and second that his justification for pursuing it lay in the chance that since his work over the years had not 'been without influence on present and future events' it might, as a matter of historical interest, be worth setting them on record.[24] Precisely when he decided to write a formal autobiography is uncertain, for his initial plans took a different slant. There is, among his papers, a typed schema for a possible book. Headed 'Landscape with Machines' it actually bears very little resemblance to the existing work of that title.[25] Instead, it sketches out what is not so much a narrative of Tom's life as a survey of his lifelong interests. He heads the opening chapter 'For the Love of the Job' and allocates subsequent chapters to his lifelong love of railways; the development of the motor car; the pleasure that motor-racing has brought him, and his engagement with the waterways. Above the plan-proper, he types 'The Wine in the Bottles. (Introduction) (quote Lowes Dickinson)'.

Goldsworthy Lowes Dickinson (1862–1932), to whom Tom's thinking apparently turned, had been a Cambridge academic – a political scientist and philosopher. By inclination a humanist and a pacifist, he belonged to the Society for Psychical Research and had connections with the Bloomsbury Group. The phrase, 'The Wine in the Bottle' on which Tom seized as a possible title, derived from his 1905 work, *A Modern Symposium* in which Dickinson argued the merits of making the best of the life in the here and now, as opposed to trusting in Christianity's pledge of a life-after-death. 'If you project, so to speak, all your goods into the future', he wrote:

> That shows that you don't appreciate those that belong to life just as it is and wherever it is. And there must, I am sure, be something wrong about a view that makes the past and the present merely a means to the future. It is as though one were to take a bottle and

turn it upside down, emptying the wine out without noticing it; and then plan how tremendously one will improve the shape of the bottle. Well, I'm not interested in the shape of bottles. And I am interested in wine.....[26]

Enjoying the passage Tom contemplated adopting Dickinson's emphasis on enjoyment of wine, as opposed to earnest endeavours to improve the design of wine bottles, as an analogy for the life well-lived.

What led him to abandon the prospect of writing a rag-bag mixture of memoir and reflection in favour of telling his life's story we do not know but the decision made sound sense. Whether, even before the full seriousness of his illness became apparent, Tom had some presentiment of his early death is a tantalising question and that he should describe it as 'an act of self-indulgence' which he never expected to perpetrate sounds rather like the special pleading of a man who believed that he was living on borrowed time.[27] He prefaced the work with the rueful verse that he saw above a cottage doorway in Aller Park when Sonia and he stayed there in 1953:

Of dying man
His living mind
By writing deeds
His children find.[28]

The idea of writing with the intention of setting out his life's works – to some degree justifying himself – to his sons while he still has the time clearly appealed to him. That he declines to set out the reasons for the breakdown of his and Angela's marriage in the book – 'such things are not to be set down in an autobiography' he maintains – and avoids any explicit reference in its pages to his often stormy relationship with Kyrle Willans suggests that he envisaged that it should be a conciliatory work, setting the seal upon a pact of peace.[29]

But his goodwill had its limits and certain individuals were for ever beyond its reach. By this time, Tom and Sonia had become friendly with Conservative politician Sir John Smith and his wife Christian. In 1964, Tom had the considerable satisfaction of co-masterminding the Great Steam Fair which Sir John instigated and which took place at his home of Shottesbrooke Park near the River Thames at Maidenhead.[30] An active member of the National Trust, Sir John came to think that the organisation did not give small historic buildings the attention that they merited and founded the Landmark Trust to make up the deficiency. Not only would it make its buildings available as holiday lets, but it also made a point of restoring them in a way that respected their original function and character. For a time, the Landmark Trust employed Sonia to furnish these properties; a job to which she brought immense flair.[31]

The Realm of Ghosts

The 1969 general election found Smith elected MP for the Cities of London and Westminster. In celebration, he invited Tom and Sonia to a reception marking the completion of the Jesus College barge's restoration and the occasion of the vessel's being towed upriver to a mooring near Maidenhead railway bridge.

As it happened, Robert Aickman was also among the guests. Amid the party atmosphere, each man deliberately avoided the other until Sonia, exasperated, made a point of engaging Robert in talk. Since he received her overture amiably, she drew Tom into their exchange with the result that they managed 'quite a good conversation'.[32] If it sounds as though the *rapprochement* was no more than tepid, it nevertheless paved the way for an exchange of letters in which Robert suggested that Tom might care to write a book examining the consequences of change and their tendency to pave the way for a lower standard of living.[33] But whether the good will amounted to anything more than surface amicability is uncertain. Raymond Russell quotes canal writer Hugh McKnight's recollection of attending the same dinner party as Tom in 1974 and hearing him observe that Robert Aickman was 'the most *evil* man' that he had ever known.[34]

There are further signs that the relations between the two men never entirely healed. Since *Landscape with Machines* came out in 1971, it seems likely that Tom was writing *Landscape with Canals* – the part of his autobiography which covers his dispute with Aickman – over 1971-2. That it would not be published until 1977 was probably on account of the publisher Allen Lane's worry lest Robert pursued an action against them for defamation. From the surviving manuscripts, it is evident that the section of the 'End of the Cut' chapter which deals with the Market Harborough Rally of Boats came in for exhaustive revision between drafting and publication. Someone, either one of Allen Lane's editorial staff or their lawyers, insisted upon removal of all references to 'Bloomsbury' – the name by which Tom chose to designate Aickman – from the text. When Tom described the moment at which *Cressy* passed Peter Scott's boat *Beatrice* with Aickman on board, he actually wrote:

> Although we tried to appear totally unconcerned, we were well aware of the hostile beady eyes of Bloomsbury peering covertly at us through the window curtains.[35]

In the published version, this sentence becomes:

> Although we tried to appear totally unconcerned, we were well aware of hostile eyes peering at us through the window curtains.'[36]

Any suggestion of beadiness or mention of Bloomsbury, the lawyers and publishers evidently decided, would turn knowing readers' thoughts immediately to Robert Aickman, and should therefore be removed from the text. Angela, incidentally, was outspoken in her contempt for their timidity. 'Aickman seems to be a fear figure of great importance', she exploded, writing to Sonia from her home in the South of France in 1976. 'He would LOVE to know this'.[37]

In the autumn of 1973, Tom took the serious step of contacting Angela, who had by this time settled in the Dordogne, to explain that he was ill. He seems to have made little of his debility and in her reply she expresses no great worry, while saying matter-of-factly that she is about sorry about his health troubles and sympathising over his most recent operation which, she says, sounds as though it must have been 'very uncomfortable'.[38] On the whole, he seems to have spent 1973 quietly, working and enjoying Industrial Archaeological meetings when he was well enough. Towards the end of the year, Manchester Polytechnic would give him an honorary Fellowship in recognition of his achievements.[39]

Early in 1974, Aidan Chambers, a prolific writer on supernatural subjects who had recently agreed to edit Barrie and Jenkins' 'Ghost Book Series', contacted Tom hoping that he might have something suitable for inclusion in its tenth volume. It was timely approach, for Tom was once more in hospital but since the two men had met in the past and discovered their shared enjoyment of ghostly literature Chambers' approach was, in a sense, a friends' reunion. Besides, the discovery almost thirty years after the appearance of *Sleep No More* that his ghost stories had caught renewed literary interest was evidently cheering for Tom. Replying from his bed, he told Chambers that he had warm recollections of their past meeting but added that he was, by now, suffering from 'almost continuous ill-health' and had no unpublished ghostly narratives to offer. But he added that he had thought out the 'idea and plot' for a new story and would, he promised, dictate it to his typist at the next opportunity. Delighted, Chambers sent him a contract soon after.[40] Sad to say, nothing came of it. Although Barrie and Jenkins' *Tenth Ghost Book* appeared in 1974, it includes no story by Tom.

Another writer of ghost stories with whom Tom corresponded at this time was Hugh Lamb. In the early 1970s, Lamb had an arrangement with publisher W. H. Allen to produce one anthology of ghost stories a year; the earliest of them was *A Tide of Terror* which appeared in 1972. He greatly admired 'Hawley Bank Foundry' which appears in his collection *A Wave of Fear* (1973) and he would be responsible for publishing two of Tom's later-life stories, namely 'The Shouting' which appears in *The Thrill of Horror* (1975) and 'The House of Vengeance' which appears in *The Taste of Fear* (1976).[41] In 1948, when Constable brought out the first edition of *Sleep No More*, neither 'The Shouting' nor 'The House of Vengeance' had come into being, although both tales appear in later editions of the collection. Lamb evidently had more curiosity concerning Tom's ghost stories than Chambers cared to show and he asked outright whether Tom's wife ever discovered 'the reason for the shouting'. Tom's answer does not survive. Nevertheless, he had returned to the literary genre which had first brought him to public attention.

By February 1974 when he came to write the final chapter of *Landscape with Figures*, Tom knew that his illness was terminal. To his many friends, the news of its seriousness was shocking. After all, Tom was in his sixties – hardly old – and Richard was barely 20 while Tim, in his teens, was still at school. Sam Clutton, a loyal friend since Tom's days in the VSCC wrote to him observing with admirable candour that 'it must be a curious sensation to be, so to speak, under starters' orders'.[42] Some weeks earlier, with a shrewd sense of the severity of Tom's condition, he had already written to Sonia asking whether he might take responsibility for payment of the regular monthly cash allowance – pocket money, in effect – that Tom made to his mother 'in some way that would not be apparent to her'.[43] Many of Tom's friends – his cousin David the painter and Mel Russell-Cooke's companion, now living in the Wiltshire village of Highworth; Dick Harper; Jack Simmons the railway historian and John Betjeman – asked to be allowed to visit him at Stanley Pontlarge. Christian Smith, Sir John's wife, no doubt expressed the thoughts of many of them when she remarked that 'Tom's boundless courage and wisdom' must have convinced the surgeon of the merits of 'being plain' with Sonia and himself.[44] While Tom did not subscribe to any mainstream faith, he nevertheless regarded himself as religious by inclination, that is to say, in the sense of rejecting materialism and professing a profound love for the natural world whose beauty he had discovered as a child growing up in the Welsh Marches. These instincts sustained him through his last weeks.

He died on 9 May 1974. Despite the distance that he professed to keep from mainstream Christianity, his long-term Talyllyn friend, Reverend Wilbert Awdry, would conduct his funeral in the church at Stanley Pontlarge and include in it the scriptural passage from the book of Ecclesiasticus which praises the work of the carpenter, the smith and the potter – craftsmen all:

> 'All these trust to their hands: and everyone is wise in his work,
> Without these, a city cannot be inhabited.'[45]

The nurture and reverence for practical skill and the art of the engineer had been at the heart of his life, and it was right that they should be celebrated at his passing.

Besides his books, Tom's legacy lies in an active heritage railway sector, a living canal network and a flourishing interest in the archaeology of the early industrial age. True, the thought of present day boaters ascending the Caen Hill locks in their leisure cruisers might not have been entirely what he had in mind when he embarked on his canal crusade, but he at least knew better than to try to tie the hands of his successors. More significantly his writing, springing as it does from the tension between his heartfelt love for the countryside and his equally powerful affection for manifestation of historic industry, has quietly swung our thinking about the industrial past. Where its traces survive, we tend to cherish and preserve them. Tom, it goes without saying, was never entirely at peace with himself but at the same time he had a remarkable capacity to engender a warm love for the landscapes, the buildings and the machinery which caught his imagination. His books embody his mix of surging affections and – as he had recognised aboard *Cressy* crossing Whixall Moss one serene morning on the Shropshire Union canal – a haunting vision of irreconcilable opposites drawn into perfect harmony.

After his death, Sonia received letters of condolence from Tom's many friends, Dick Harper, Barbara Willans, and Jack Simmons among them.[46] Betjeman sent her a copy of the sketch that he had made some twenty years previously of Stanley Pontlarge in the guise of a stout, cloth-capped labourer with hobnail boots and an ashplant walking stick.[47] There was a distraught message from Angela. 'I cannot explain to you how upset and sad I am', she wrote. 'Tom will be so greatly missed by so many people including me' and she ended helplessly, 'I cannot say anything adequate...all my sympathy goes to you.'[48] Robert sent a snide note, complaining about Tom's treatment of him while at the same time saying how much he would have liked the two of them to be reconciled.[49]

Cherishing his legacy, Sonia worked assiduously to ensure that his books remained in print. It was not an easy task, particularly since publishing was changing fast. Early in 1975, for instance, Penguin Books wrote to Hilary Rubinstein at A. P. Watt, the literary agency which handled Tom's books after his death, to announce their decision to reduce the royalty payable on *Isambard Kingdom Brunel* from 15% to 10% in order to keep the cover price competitive. By way of explanation Peter Carson, Penguin's editor-in-chief, wrote to Sonia to say that the firm's decision was 'indicative of current trends in publishing'.[50] Unimpressed, she observed to Rubinstein that it seemed to her that publishers had 'done well in enough in the good times to have rather more fat to live off than the majority of authors'. At the same time, feeling herself caught between a rock and a hard place, she accepted 'the inevitable' and agreed to take a twelve-and-a-half per cent royalty 'on the hardback Brunel'.[51] She continued to support both the Talyllyn Railway and the Inland Waterways Association throughout her life and was given honorary life membership of a number of organisations with which Tom had been associated, among them, the VSCC, the Bugatti Owners Club and the Newcomen Society on whose Council she served – the first woman ever to do so.[52] Her services to industrial archaeology and heritage led to the award of an OBE in 2011.

In time, and with so prolific an author, it was hardly surprising some of Tom's titles went out of print – *Horseless Carriage* and *Lines of Character* among them – and are now available only through the second-hand market. But others, notably *Narrow Boat* which was reissued for the seventieth anniversary of its publication, and *Red for Danger* and the lives of Brunel, Telford and the Stephensons remain widely available and continue to sell.[53] In 2020, interviewed by Harriett Gilbert on BBC Radio Four's eponymous programme, Radio DJ Stuart Maconie made *Sleep No More* his choice of 'A Good Read'. The History Press have re-issued Tom's history of early flight – *The Aeronauts* – albeit changing the book's enigmatic, daredevil title to *The Balloonists* which is more literal, if rather dull. *Winterstoke*, for which Tom almost despaired, found a new lease of life in the 'Faber Finds' series while *High Horse Riderless* caught the attention of environmentally concerned readers in the 1980s and has been reissued as a 'Green Classic'.

As for Tom's lasting legacy, he was broadly inclined to share Richard Trevithick's view that 'the great honour of being a useful subject' counted for more than either riches or fame.[54] It was, his obituarist John Guest wrote, his ambition 'to give the history of the industrial revolution an imaginative and literary shape'.[55] Although a number of his books are no longer in print, their achievement in shaping the way that we view the past lives on, most notably in the value that we have come to ascribe to industrial history. Brunel, for example, whose significance largely escaped the 1950s publishing establishment, is now ensconced in the primary Key Stage Two syllabus. That an All Party Parliamentary Group was established in 2015 to support British Industrial Heritage Sites would probably amuse Tom considerably. More significantly, it might please him to reflect that any of us who mark the grooves which towing lines have worn in the side of canal bridges, or follow the overgrown tracks of half-forgotten tramroads, or find beauty in industrial buildings all bearing what he once called the 'patina of use' follow in his steps and stand within his shadow.

Notes

Abbreviations

General
BMD [Records of birth, marriage and death registrations] – www.freebmd.org.uk
IGMT – Ironbridge Gorge Museums Trust

Books by L.T.C. Rolt
LoC – L.T.C. Rolt, *Lines of Character* (Constable, London, 1952)
LWM – L.T.C. Rolt, *Landscape with Machines* (Alan Sutton, Gloucester 1984; First published Longmans, London, 1971)
LWC – L.T.C. Rolt, *Landscape with Canals* (Alan Sutton, Gloucester 1984; First published, Allen Lane, London, 1977)
LWF – L.T.C. Rolt, *Landscape with Figures* (Alan Sutton, Gloucester 1992)
RfD – L.T.C. Rolt, *Red for Danger*. (4th edn., David & Charles, Newton Abbot, 1982; First published, John Lane, The Bodley Head, Ltd, London, 1955)

Books by other authors
CY – Ian Mackersey, *Tom Rolt and the Cressy Years* (M. & M. Baldwin, London, 1985.)
TRRU – Robert Aickman, *The River Runs Uphill* (J.M. Pearson, Burton upon Trent, 1986; posthumous publication)

Chapter One

1. LWM p.28.
2. LWM p.27.
3. L.T.C. Rolt, *Green and Silver* (George Allen & Unwin, Ltd, London, 1949) pp.22–3.

4. Colonel Sir John Rolt, *On Moral Command* (W. Clowes, London, 1842) p.42.
5. Ibid, p.38.
6. Ross, Sir John Foster George, of Bladensburg, *A History of the Coldstream Guards* (A.D. Innes & Co., London 1896) p.469; *London Evening Standard*, 5 May 1853, notice in the 'Marriages' column.
7. Entry for Edith Elizabeth Rolt, c.1854–1880 in www.thepeerage.com; *The Globe*, 19 February 1858.
8. LWM p.6.
9. *Cheltenham Journal and Gloucestershire Fashionable Weekly Gazette*, 7 January 1865.
10. LWM p.7.
11. *Cheltenham Chronicle*, 12 November 1887.
12. *Gloucestershire Echo*, 8 December 1887.
13. LWM p.6.
14. LWM p.16.
15. LWM pp.9–10.
16. www.christleton.org.uk/christleton3/2018/history2018/lionelgarnett; *Morning Chronicle*, 20 September 1859.
17. *Cheshire Observer*, 11 November 1899 and 21 April 1900.
18. *London Evening Standard*, 2 April 1896.
19. *Manchester Courier and Lancashire General Advertiser*, 16 October 1880; also LWM p.4.
20. Advertisement of sale of the contents of Cogshall Hall on 'instructions from the Executors of Thomas Clarke, esq., deceased' in the *Chester Chronicle*, 17 April 1886.
21. *Reading Mercury*, 14 May 1887 which mistakenly gives Timperley's initials as 'T.R' rather than 'I.R'.
22. The entry for Herbert Luard Timperley on www.unithistories.com/Army_officers gave the year of his birth as '1888–?'. Unfortunately, it appears that this website is no longer extant.
23. For Rev. I.R. Timperley's Oulton appointment see *Manchester Courier and Lancashire General Advertiser*, 12 November 1891; for his transfer to Sutton Guilden see *London Evening Standard*, 2 April 1896.
24. *Morning Post*, 30 November 1905 and https://southwark.anglican.org/brief-history-of-the-diocese.
25. *Cheshire Observer*, 17 February 1906.
26. LWM p.6. In January 1940, reservist Herbert Luard Timperley, who was now aged over 50, received an emergency commission in the Royal Regiment of Artillery and transferred to the Pioneer Corps a month later. Captured near Abbeville in the early summer of 1940, in June he was reported as 'missing'. A month later, it emerged that he been taken prisoner

and transported to Poland; he was eventually repatriated in November 1943. (Information from www.forces-war-records.co.uk.)
27. Thomas Clarke of Cogshall Hall made his will in January 1885; he died on 17 February 1886. The will was proved in London and probate granted on 27 March 1886. https://probatesearch.service.gov.uk/#wills.
28. LWM p.6.
29. RfD p.95. Tom recounts the disaster in detail and analyses its causes.
30. LWM p.4.
31. LWM p.9.
32. Entries for N.C. Rolt and W.F.C. Rolt on www.historyofceylontea.com/tea-planters/planters-registry.
33. Rolt, *Green and Silver* (1949) p.23.
34. Lily Rolt's entry in the 1911 UK Census. (My thanks to Julian Hunt of the Bucks Archaeology Society for directing me to this information.)
35. *Cheshire Observer*, 8 April 1911.
36. *Chester Chronicle*, 28 November 1914.
37. *Brecon County Times*, 2 October 1913.
38. *Brecon County Times*, 12 March 1914. Tom suggests that the house dated from the turn of the century (LWM, p.14).
39. LWM pp.24–5, where Tom gives Llydyadyway as 'Lidiart –y-Wain'. Between 1910 and 1920, the *Brecon County Times* makes frequent mention of the Lilwall family.
40. LWM p.25.
41. IGMT Rolt Collection, ref R 1/2/04.
42. LWM p.22.
43. *Brecon County Times*, 15 February 1915 and 9 September 1915.
44. IGMT Rolt Collection, ref R 1/2/04.
45. IGMT Rolt Collection, ref R 1/2/01.
46. Ibid.
47. LWM p.15.
48. LWM pp.15 and 27.
49. LWM facing p.37.
50. LWM pp.57 and 39.
51. LWM p.23.
52. *London Gazette* (Supplement) 28 June 1916.
53. LWM pp.22–23.
54. LWM p.18.
55. LWM p.26.
56. LWM p.21 and photograph facing p.37.
57. LTC Rolt, 'Cwm Garon' in *Sleep No More*. (History Press: Stroud, 2010) pp.33–49; First published Constable, London, 1948.)
58. LWM p.17.
59. LWM pp.11 and 25.

Notes

Chapter Two

1. *England's Overland Mail,* 22 December 1904, pp.17–20.
2. *Civil and Military Gazette (Lahore),* 17 March 1907.
3. *Homeward Mail from India, China and the East,* 22 May 1909, p.9.
4. *Cheshire Observer,* 18 June 1910, which notes Algernon's engaging Hon Arthur Lyulph Stanley, sometime MP for Knutsford, in vigorous debate about the widespread unemployment of the time.
5. LWM p.9.
6. LWM p.47 and 1922 advertisement for Belsize-Bradshaw exhibited at the Olympia Motor Show, Stand 289. An image of it appears at www.gracesguide.co.uk/File:Im19221103CyCar-belsize4.jpg.
7. LWM p.43.
8. The *Cheltenham Chronicle,* 30 April 1921 carries a notice from Messrs Engall, Cox & Co announcing that the Cottage, Stanley Pontlarge, is to be sold at auction on 19 May 1921.
9. *Illustrated Sport and Dramatic News,* 23 March, 1907; *Surrey Mirror,* 26 May 1911.
10. *Hampshire Chronicle,* 30 May 1908 and www.bhcgodalming.org/church-history/previous-rectors.
11. *Surrey Advertiser,* 28 November 1914; *Gloucestershire Echo,* 17 June 1932.
12. *Gloucestershire Echo,* 17 June 1932; *Gloucester Citizen,* 1 July 1932.
13. *Gloucestershire Echo,* 1 July 1932.
14. LWM p.51.
15. LWM p.44.
16. LWM pp 28–9.
17. LWM pp.33–4. Designed by Francis Webb, the LNWR 'Precedents' were 0-2-4 locomotives intended for use with express passenger trains. Introduced between 1910 and 1915, the LNWR George the Fifth Class were 4-4-0 passenger steam locomotives designed by Charles Bowen-Cooke. The LNWR Precursor class included locomotives of three distinct types: 2-4-0 locomotives designed by Webb in the 1870s; 4-4-0 locomotives designed by George Whale in the 1900s; and a class of 4-4-2 tank engines, also of Whale's design and dating from 1906–09. (Wikipedia.)
18. LoC p.102.
19. LWM pp 38–9 and photograph facing p.52.
20. LWM p.46.
21. 'The Railway, being a collection of Railway Stories, Verses and Anecdotes by L.T.C. Rolt', IGMT Rolt Collection, ref R 1/4A/03.
22. LWM pp.22 and 48.
23. *Gloucestershire Echo,* 3 January 1922.
24. LWM pp.48–50; *Brecon County Times,* 12 January 1922; Wikipedia entry for Herbert Rowse Armstrong.

25. LWM p.49.
26. LWM p.50; *Gloucestershire Echo*, 8 April 1922.
27. Tom was in no doubt about Armstrong's guilt but there has been some recent suggestion that he may have been 'the victim of a frame-up'. (J.S. Barnes, 'A Hanging Offence', *Times Literary Supplement* 8 July 2022, p.26, reviewing Stephen Bates, *The Poisonous Solicitor,* Icon Books, London, 2022.)
28. LWM p.25.
29. LWM p.40.
30. LWM pp.39–41.
31. LWM p.26, where Tom quotes Thomas Traherne, *Centuries of Meditation*, 'The Third Century', Section 3.
32. LWM p.40.
33. Cheltenham College Record Books, 1922–23 et seq. and L.T.C. Rolt's Cheltenham College Record Card.
34. Ibid.
35. Joseph Boughey's email to the author concerning his exchange with the late Sonia Rolt.
36. LWM p.42.
37. Cheltenham College Record Books, 1922-26 and L.T.C. Rolt's record card.
38. LWM p.41.
39. LWC p.32 and Cheltenham College registers.
40. Warner, George Townsend and C.H.K. Marten, *The Groundwork of British History* (1923); Burrell-Smith, G, *Outlines of European History, 1789-1914* (1920) and Unstead, J.F. and E.G.R. Taylor, *Essentials of World Geography* (1916).
41. Cheltenham College Record Card for L.T.C. Rolt, entry for term 1924.
42. LWM p.203.
43. Cheltenham College Report Card for L.T.C. Rolt, entries for 1924-1926.
44. LWM p.57.
45. Cheltenham College Record Book 1925–6.
46. Cheltenham College Report Card for L.T.C Rolt, entry 1926 For Rolt's recollections of the interview, see LWM p.58. One of Hardy's obituarists observed that while he was a 'strict disciplinarian, [...] a warm humanity' lurked 'beneath the strictness'. (*Birmingham Daily Post,* 27 December 1958.)

Chapter Three

1. Jemima Clarke and Hero Taylor were friends before either of them married. In 1910, Hero married Kyrle Willans; her brother George Taylor married Jemima's sister Augusta Clarke and Jemima married Lionel Rolt. Connections between the three families were, therefore, strong.
2. www.gracesguide.co.uk/Peter_William_Willans.

Notes

3. *Evesham Standard and West Midland Observer* 15 October 1921 and 3 June 1922.
4. Theo Sherwen, *The Bomford Story* (Bomford & Evershed Ltd., Evesham: 1978) p.10 and www.bomford.net.
5. www.bomford.net/worcestershire and LWM p.68.
6. LWM p.67.
7. Dr Bruce Bomford, *The Bomfords of Worcestershire* (Private publication, 1983) on www.bomford.net, section headed 'Steam Ploughs'.
8. LWM p.70.
9. LWM p.71.
10. LWM p.72.
11. LWM p.65.
12. LWM p.75.
13. LWM p.76.
14. LWM pp.66 and 74.
15. LWM p.75.
16. LWM p.73.
17. LWM p.74.
18. Dr Bruce Bomford, *The Bomfords of Worcestershire* (Private publication, 1983) on www.bomford.net, section headed 'Steam Ploughs'.
19. LWM pp.62 and 70.
20. LWM p.70.
21. LWM p.73; RfD p.69.
22. *Lincolnshire Standard & Boston Guardian*, 17 November 1923.
23. LWM pp.78–9.
24. LWM p.76.
25. D.R. Bomford, *Corn in England*, (Private publication, 1927) pp.5–7.
26. LWM p.81.
27. See illustration 5. Another picture of Bomford's ploughing engine appears in Theo Sherwen, *The Bomford Story* (Evesham: Bomford & Evershed Ltd, 1978) p.37.
28. LWM p.81 and www.railway-technical.com/glossary/steam-glossary.html (entry for 'Fusible Plug').
29. LWM p.82.
30. Ibid and Theo Sherwen, *The Bomford Story* (Evesham: Bomford & Evershed Ltd, 1978) p.39. The Gyrotiller – an early form of power harrow – was the invention of one Norman Storey. It ran on caterpillar tracks and had mechanised tines or 'skives' to break up and level the ground at its rear. John Fowler & Co of Leeds licensed the rights to build it in 1925 (www.gracesgruide.co.uk).
31. LWM p.68.
32. LWM p.69 and www.gracesguide.co.uk/Kyrle_William_Willans.
33. Details of the Sentinel Rhinoceros from the Farmers' Weekly website: www.fwi.co.uk/machinery/tractors/machinery-milestones-steam-powered-tractors.

34. www.sentineldriversclub.com/history_1923-28.
35. Dr Bruce Bomford, www.Bomford.net 'Bevington' section.
36. Writing in the early 1970s, Tom suggests that Willans 'would now be called [Kerr Stuart's] chief development engineer' (LWM p.92).
37. LWM p.92.
38. LWM p.85; letter from Messrs. Kerr Stuart to Tom of 10 April 1928 noting his wish to start working for them on 1 May 1928. IGMT Rolt Collection, ref R 2/1F/01.
39. Douglas Bomford, undated testimonial for LTC Rolt. IGMT Rolt Collection, ref ROLT 02/1F/01 (1)

Chapter Four

1. LWM pp.85–6.
2. LWM p.85.
3. LWM p.89.
4. LWM pp. 92–3.
5. The 'check' was a brass disc. Kerr Stuart employees handed their 'checks' in at the works' office each morning to clock on; the 'check boy' returned them to their owners during the day.
6. LWM p.93.
7. LWM p.86.
8. LWM pp.86–7.
9. Bomford describes himself as 'Works Manager' in a testimonial that he wrote for Tom after Kerr Stuart's collapse. IGMT Rolt Collection, ref ROLT 02/1F/01 (3).
10. LWM p.98.
11. Two manuscript versions of 'Strange Vista' survive; one finished, one unfinished. Both are in the possession of the Rolt family. Neither has appeared in print.
12. L.T.C. Rolt, 'Strange Vista' (unpublished mss, courtesy of Timothy Rolt).
13. 'Hawley Bank Foundry' in L.T.C. Rolt, *Sleep No More* (History Press: Stroud, 2010; First published 1948) pp.83–99.
14. *Staffordshire Sentinel*, 18 March 1929.
15. LWM p.99.
16. *Western Morning News*, 31 August 1928.
17. LWM p.100.
18. LWM p.102.
19. https://ks4415.blogspot.com/p/ks-4415-early-history.html; and https://ks4415.blogspot.com/p/the-trials-at-dinas-15th-november-1928.html.
20. *Staffordshire Sentinel*, 15 November 1928. Extracts from the other journals appear on https://ks4415.blogspot.com/p/the-trials-at-dinas-15th-november-1928.html.

Notes

21. See: https://www.festipedia.org.uk/wiki/Kerr_Stuart_4415, citing B. Webb, *The British Internal Combustion Locomotive: 1894 – 1940.* (Newton Abbot: David & Charles, 1973) pp.60–61.
22. L.T.C. Rolt, *A Hunslet Hundred* (David and Charles: Dawlish, 1964) pp.100–101. See also www.kerrstuart4415.org.uk/history.
23. *Staffordshire Sentinel*, 14 October 1929. A 'light railway' is one that is purposely built to less stringent engineering standards than a normal railway. The tracks may be unfenced; the gradients steeper and the curves sharper. To reduce costs and facilitate construction, the gauge may well be narrow and the signalling system simple. The maximum permitted speed will be low. Most UK light railways were constructed under the provisions of the 1896 Light Railways Act, although the term also covers similar lines built before the legislation's enactment. (Information from Alan A Jackson, *The Railway Dictionary*, (Alan Sutton: Stroud, 1992.)
24. Tom's essay is in IGMT's Rolt Collection, ref ROLT 01/4A/12b.
25. *Welsh Mountain Railways* (London: GWR, 1924.) No author named.
26. LWM p.105.
27. LWM p.109.
28. Compare LWM p.109 with RfD p.93.
29. LWM p.61.
30. LWM pp.110–11. Tom suggests that *Cressy* took her name from HMS *Cressy*, an armoured Chatham Division cruiser. Built in 1899, HMS *Cressy* was torpedoed by a German submarine on 22 September 1914. https://thedockyard.co.uk/the-collections/digital-exhibitions/three-cruisers/three-cruisers_sailors/.
31. LWM p.112.
32. LWM p.92.
33. LWM p.112.
34. LWM pp.110–3.
35. LWM p.167.
36. LWM p.113.
37. LWM p.114.
38. LWM p.103.
39. The case of the Peninsular Locomotive Company v H. Langham Reed would come before the Patna High Court in 1936. (https://www.casemine.com/judgement/in/56e0f278607dba38965fcb32.)
40. The *Yorkshire Post and Leeds Intelligencer*, 15 February 1929 gives a full account of Kerr Stuart's 1929 AGM.
41. A Catalogue for Evos Doorways Limited (undated, but probably late 1920s–c.1930) is available at: https://archive.org/details/DoorsAndDoorways.
42. *Daily Herald*, 31 August, 7 September, 11 September 1929; *Dundee Courier*, 12 December 1929.

43. *Yorkshire Post and Leeds Intelligencer* 24 March 1930 and LWM p.132, which calls the company 'Evos Sliding Doorways Ltd'.
44. LWM p.132.
45. *Staffordshire Sentinel*, 24 March 1930.
46. *Staffordshire Sentinel*, 9 May 1930.
47. *Staffordshire Sentinel*, 10 May 1930.
48. *Staffordshire Sentinel*, 12 May 1930.
49. *Staffordshire Sentinel*, 15 May and 2 June 1930.
50. *Staffordshire Sentinel*, 21 June 1930.
51. *Staffordshire Sentinel*, 15 July 1930.
52. LWM pp.130–131. Tom dates this epic journey to July 1930.
53. *Staffordshire Sentinel*, 14 August 1930.
54. *Staffordshire Sentinel*, 15 August 1930.
55. Apparently the Order was issued on 14 October 1930. (*Staffordshire Sentinel*, 15 November 1930.)
56. *Staffordshire Sentinel*, 14 November 1930.
57. *Staffordshire Sentinel*, 18 November 1930.
58. *Staffordshire Sentinel*, 17 November 1930.
59. LWM p.132.
60. Ibid.
61. *Scotsman*, 27 November 1930.
62. *Daily Mirror*, 6 November 1935.
63. LWM p.132.
64. *Western Mail,* 12 January 1934; Scott Murray, *The Title – the Story of the First Division* (Bloomsbury, 2017) p.96.
65. www.gracesguide.co.uk/Hunslet_Engine_Co and L.T.C. Rolt, *A Hunslet Hundred*, p.101.
66. When Talyllyn locomotive No. 4, *Edward Thomas,* one of Kerr Stuart's 'Tattoo' class, required large-scale boiler repair, Mr John Alcock, Talyllyn member and Hunslet's Managing Director, offered to do the work required at no cost to the Talyllyn Company. L.T.C Rolt, *Railway Adventure* (London, Constable, 1953) p.141.

Chapter Five

1. www.dursleyglos.org.uk and Wikipedia entry for Margaretenhohë.
2. *Gloucester Citizen,* 22 April 1931.
3. LWM p.147.
4. Ibid.
5. Definition from www.investopedia.com.

Notes

6. LWM p.146.
7. LWM p.147.
8. www.dursleyglos.org.uk/html/dursley/industry/listers/family/lister_family.htm.
9. *Gloucester Citizen*, 3 December 1931; *Birmingham Daily Gazette*, 4 December 1931; *The Scotsman*, 4 December 1931.
10. LWM p.137.
11. 'Situations Vacant' column of the *Gloucester Citizen*, 13 and 16 September 1930, in which Mrs Vanderguilt advertised for housemaids at The Towers.
12. J. R.S Malloch, known as Ron, was the son of Robert Anthony Malloch, chairman and managing director of Dangar, Gedye and Malloch, Ltd, the Sydney-based Agricultural Engineering firm who were Listers' New South Wales agents and distributors. (http://oa.anu.edu.au/obituary/malloch-robert-anthony-663J.)
13. LWM p.136
14. LWM p.118.
15. LWM p.143.
16. LWM p.136.
17. LWM p.140.
18. LWM pp.139–141
19. LWM p.148.
20. LWM p.151.
21. LWM p.167. For the various posts that Tom held at this time, see LWM pp.151–67.
22. LWM p.158 and Brighton Run programme, IGMT Rolt Collection, ref ROLT 13/3/4/1-gen-eve/02/.
23. LWM pp.168; 202–3.
24. IGMT Rolt Collection, ref R1/4C/1.
25. George Frederic Stewart Bowles (1877–1955) the author of the poem, was a barrister, MP for Norwood (1906–1910) and author of *The Strength of England* (Methuen, 1926). His father was Thomas Bowles, founder of *The Lady* magazine. Exactly when Bowles wrote the poem is unknown although – for what it is worth – *The Carriage Journal*, vol. 15, No. 1, p.235 (Summer 1977) quotes the full version. It also appeared in an early (1903) issue of *The Car* magazine. (Information from *MotorSport*, 7 July, 2014.)
26. L.T.C Rolt, 'Strange Vista', (unpublished novel). Red Notebook manuscript in the collection of Timothy Rolt.
27. LWM p.170.
28. LWM p.174. An affectionate obituary of Carson appears in the October 1989 issue of *MotorSport* Magazine: www.motorsportmagazine.com/archive/article/october-1989/6/around-and-about-october-1989.

29. LWM pp.174–77.
30. John Minnis, 'Defining the Vintage Car: L.T.C. Rolt, the vintage sports car and the rise of motoring heritage' in *Industrial Archaeology Review*, 39, May 2017, pp.2–13 at p.3.
31. Minnis, (2017) p.3 and Minutes of the VSCC Committee, 21 February 1935.
32. LWM p.176 and Minnis (2017) p.3.
33. LWM p.179.
34. LWM pp.158 and 180.
35. LWM p.180.
36. VSCC Minutes, 15 January 1936 and Minnis (2017) p.5.
37. Minnis (2017) p.5.
38. Minutes of VSCC AGM, 15 January, 1936. VSCC Minute Book 1, VSCC Library.
39. VSCC Committee Minutes, 11 September 1935.
40. For Tom's committee-level contributions to the VSCC see Minnis (2017) p.4.
41. LWM pp.143, 174–5, 181 and 193.
42. LWM pp.185 and 193.
43. LWM pp.193–4.
44. LWM, p.194.
45. LWM pp.194–6 and bundle of letters dated 1937–1938 (from Gloucestershire Dairy Company, Sam Clutton and Bugatti Owners' Club et al) regarding the acquisition of Prescott Hill near Cheltenham for hill trials. IGMT Rolt Collection, ref ROLT 06/2/BOC/01.
46. LWM p.218 and Victoria Owens, 'L.T.C. Rolt's Stories of the Supernatural – *Sleep No More* (1948)' in *Western Courier* 25 March 2020 (Newsletter of the Western Region of the Newcomen Society) pp.7–10. 'New Corner' first appeared in *Mystery Stories*, issue 27.

Chapter Six

1. LWM pp.118, 143 and 160.
2. Cara's surviving correspondence with Tom is in IGMT's Rolt Collection, ref ROLT 01/5A/03. Neither of these letters carries a date.
3. LWM p.204.
4. LWM pp.207–9.
5. LWM p.206.
6. Telegram, Cara to Tom, 28 November 1935. IGMT Rolt Collection, ref ROLT 01/5A/03.

Notes

7. Cara to Tom, 30 December 1935, Odiham. IGMT Rolt Collection, ref ROLT 01/5A/03.
8. Cara to Tom, 14 January, no year given. IGMT Rolt Collection, ref ROLT 01/5A/03.
9. Cara to Tom, 14 January [1936?]. IGMT Rolt Collection, ref ROLT 01/5A/03.
10. Cara to Tom, 22 January 1936, from 'The Vacant Cottage'. IGMT Rolt Collection, ref ROLT 01/5A/03.
11. LWM pp.169 and 213.
12. LWM p.207 and Cara to Tom from 56 Trinity Court, WC1, no date. IGMT Rolt Collection, ref ROLT 01/5A/03.
13. LWM, pp.210–212.
14. Cara to Tom, nd., given address 'Odiham'. IGMT Rolt Collection, ref ROLT 01/5A/03.
15. Cara to Tom, nd., given address 'The Orchard'. IGMT Rolt Collection, ref ROLT 01/5A/03.
16. Cara to Tom, nd., given address, 'The Orchard, Windlesham'. IGMT Rolt Collection, ref ROLT 01/5A/03.
17. LWM pp.213–4.
18. LWM p.199.
19. LWM p.251, quoted in CY p.18.
20. L.T.C. Rolt, *Narrow Boat* (Eyre and Spottiswoode, London, 1944) p.14.
21. The *Leicester Chronicle*, 21 May 1932 includes a picture of Mr Fortune and his bride Mollie Stubbings setting off from Belgrave Boathouse in what looks like a motor launch for their honeymoon. In June 1934, Mollie Fortune broadcast about a 'dinghy trip' that she had made on the Trent from Nottingham to Swarkeston Bridge near Derby, which is indicative both of her enjoyment of boating and her gifts as a radio-journalist (*Nottingham Evening Post*, 8 June 1934).
22. LWM p.215.
23. LWM p.217.
24. VSCC committee minutes of meeting held on 26 August 1937.
25. *Gloucestershire Echo*, 14 September 1925.
26. For example, *Gloucester Citizen*, 7 November 1930 (funeral of Herbert Wykeham-Musgrave, JP*)*; *Gloucester Citizen*, 16 January 1932 (Cirencester Hospital Ball); and *Western Daily Press*, 23 January 1932 (Vale of White Horse Hunt Ball).
27. http://www.motorsportmemorial.org/; and *Tatler*, 23 May 1934, p.4.
28. LWM p.219.
29. Ibid.
30. LWC p.14 and *MotorSport Magazine*, December 1937, p.34.

31. LWM p.219. Based in Bath, Sidney Horstmann's company, 'Horstmann Cars' was in production from 1913 to 1929. The firm dropped the second 'n' from their name in about 1920.
32. LWM p.220.
33. LWM p.226.
34. LWM p.218 and drawings of mb *Wanderer*.
35. LWM p.218.
36. LWM p.220.
37. LWM p.221.
38. LWM pp.221–2.
39. VSCC Bulletin vol. 4, No. 3, September 1938.
40. Tim Carson, who joined the RAF in the Second World War, offered to store both cars at the Sarum Hill Garage in Basingstoke which he ran. In the war years, the Horstmann mysteriously vanished, but the Humber survived. (LWM, p.222 and Clive Windsor-Richard's memoir of Tim Carson at www.vscc.co.uk/vsccMedia/27136.jpg.)
41. LWM pp.222–223.
42. Email from Will Willans (Bill's son) to the author, June 2023, referring to his father's notes on LWM.
43. *Chester Courant and Advertiser for North Wales*, 6 December 1905 (National Library of Wales, online Welsh Newspapers https://newspapers.library.wales/view/3807239/3807246).
44. *The Bystander*, 10 June 1908.
45. *Portsmouth Evening News*, 2 October 1930.
46. LWM pp. 227–229.
47. Ian Mackersey met Angela Rolt in France during the early 1980s and found her extremely frank on the subject of her parents and their treatment of Tom. CY, p.23.
48. Advertising for a housemaid, Mrs Orred gives her London address as 41 Ovington Square. *Western Gazette*, 13 July 1928.
49. *North Wiltshire Herald*, 20 March 1936; *Bath Chronicle* 20 February 1937; *Gloucestershire Echo*, 2 November 1937; *Reading Mercury* 25 November 1939.
50. CY p.23.
51. LWM p.228.
52. CY p.22.
53. LTC Rolt, *Narrow Boat* (London: Eyre & Spottiswoode, 1944) pp.28–9.
54. LWM, p.229.
55. LTC Rolt, *Narrow Boat* (London: Eyre & Spottiswoode, 1944) p.46.
56. CY p.23; LWM p.229.
57. *Reading Mercury*, 29 July 1939.
58. LWM p.183.

Chapter Seven

1. IGMT Rolt Collection, ref ROLT 01/6A/01.
2. '*Cressy* log', 29 November 1939 and *Narrowboat* (1944), p.175. Harry Rogers, coracle maker of the Severn, told Tom a similar story (LWC, p.57).
3. '*Cressy* log', 23 August, 25 August and 3 December 1939.
4. '*Cressy* log', 2 August 1939.
5. CY p.28.
6. H.R. de Salis, *Bradshaw's Canals and Navigable Rivers of England and Wales – A Handbook of Navigation* (London: Henry Blacklock & Co Ltd, 1904) Glossary of Canal Terms, pp.470–80.
7. '*Cressy* log', 10 August 1939.
8. '*Cressy* log', 11 August 1939.
9. *Leicester Daily Mercury*, 12 August 1939.
10. '*Cressy* log', 13 August 1939.
11. LTC Rolt, *Narrow Boat* (London: Eyre & Spottiswoode, 1944) p.83.
12. '*Cressy* log', 14 August 1939 and *Narrow Boat* p.86.
13. LWC p.14.
14. Ibid.
15. *Daily News (London)* 14 August 1939, et al.
16. *Western Daily Gazette*, 25 August 1939.
17. '*Cressy* log', 21 August 1939. Tom recovered one of the windlasses, having cycled back to the lock; the other must, he thought, have been appropriated by a passing boat crew.
18. '*Cressy* log', 1 September 1939.
19. *Staffordshire Sentinel*, 28 August 1939.
20. CY, p.31.
21. LWC, p.16.
22. 'This Modern Generation – a reply to its critics.' IGMT Rolt Collection, ref ROLT 01/2/06.
23. www.nationaltransporttrust.org.uk/heritage-sites/heritage-detail/rolls-royce-factory-crewe; *Crewe Chronicle*, 22 July 1939.
24. LWC p.17.
25. LWC p.18.
26. *Crewe Chronicle*, 7 October and 4 November 1939.
27. LWC p.18.
28. '*Cressy* log', 3 November 1939.
29. LWC p.19.
30. In LWC, Tom claims that they 'set sail from Church Minshull' on 15 November 1939, the date when, according to the *Cressy* log, they actually travelled some four miles and a furlong from Market Drayton.
31. '*Cressy* log', 6 December 1939.

32. LWM p.202. For Harold John Massingham (1888–1952) see https://en.wikipedia.org/wiki/H._J._Massingham.
33. H.J. Massingham, *Shepherd's Country* (Chapman & Hall: London, 1938) p.152.
34. Ibid, p.195.
35. Ibid, p.201.
36. '*Cressy* log', 2 August 1939.
37. LTC Rolt, *Narrow Boat* (London: Eyre & Spottiswoode, 1944) p.72.
38. CY, pp.34–5.
39. The '*Cressy* log' mentions 'Locks 30 and 31, Fowles and Brunsden'. The numbering has changed since the Rolts' journey – Brunsden Lock is now No. 77 and so, perhaps, have the names. 'Fowles Lock' has proved impossible to trace, but one Reverend Thomas Fowle, a past clergyman of Kintbury, may have lent his name to what is now Kintbury Lock No. 78. (https://janeausten.co.uk/down-the-kennet-and-avon-canal-with-Jane-Austen.)
40. LWC pp.26–9.
41. LWC p.25.
42. The ms is in IGMT's Rolt Collection, ref ROLT 4/01/01.
43. Philip Unwin (for Messrs George Allen & Unwin) to LTC Rolt, 21 February and 31 March 1941; Harry Batsford to LTC Rolt, 1 April 1941; E.F Bozman (for Messrs J.M. Dent) to LTC Rolt, 26 August 1942. IGMT Rolt collection, ref ROLT 04/01/05 (7).
44. Rolt, 'Imagination and Dramatic Art', in *Kingdom Come – the Magazine of Wartime Oxford* (Summer 1941, pp.113–4); 'Ethics of the Quality Car', in *MotorSport* (March 1945, pp.45–7); and 'Planned tyranny', in *Horizon* (January 1943, pp 1–5).
45. LWC p.32; CY p.42.
46. LWC p.35.
47. LWC p.52.
48. LWC p.36 and BMD.
49. '*Cressy* log', 30 August – 4 September 1941.
50. LWF pp.3–5.
51. LWC, p.53.
52. Angela Rolt to Ray Aickman [Edith Gregorson Ray] 2 November 1945 (National Archives, PRO 30/82/1).
53. IGMT Rolt Collection, ref ROLT 01/5B/01 (4).
54. LWC p.72.
55. Ibid.

Chapter Eight

1. LWC p.74.
2. https://en.wikipedia.org/wiki/Painted_Boats.

3. LWC p.84.
4. Aickman to Rolt, 9 July 1945, National Archives, Kew, Aickman papers, Ref PRO 30/82/1.
5. Ray Aickman to Joan Coster, 17 August 1945, National Archives, Kew, Aickman papers, Ref PRO 30/82/1.
6. R.B. Russell, *Robert Aickman: An Attempted Biography* (Tartarus Press: Leyburn, 2022) pp.62–3.
7. Ibid, p.53.
8. TRRU p.12.
9. Ibid, pp.14–15.
10. TRRU pp.53–4.
11. Hadfield to Rolt, 20 January 1946, letter sent on to Aickman with accompanying note by Tom dated 23 January 1946. (National Archives, Kew, ref PRO 30/82/1.) Appears also in a set of comb-bound photocopies of letters from Tom Rolt to Robert Aickman, in IGMT's Rolt Collection. They are bound in date order in separate binders for 1945, 1946, 1947, 1948, 1949 and 1950, ref ROLT 06/2/IWA/28 (1-6).
12. Rolt to Hadfield, 6 February 1946, National Archives, Aickman papers, Kew, ref PRO 30/82/2.
13. Entry for Frank Eyre in the *Australian Dictionary of Biography*, available online at https://adb.anu.edu.au/.
14. TRRU p.19.
15. LWC p.93.
16. 'The Future of the Waterways' (Inland Waterways Association, nd.).
17. CY p.65. Whether either Tom or the Tooleys appreciated the seriousness of *Cressy*'s condition is uncertain, but Mackersey notes that at this time Tom was already considering the idea of buying a more compact boat, suitable for travel over the northern canals with their short locks and in the Fens. In the event, this plan came to nothing.
18. TRRU p.16.
19. Ibid pp.16-18 and '*Cressy* log', 22 April 1946,
20. Rolt to Aickman, 25 November 1945, National Archives, Aickman papers. Kew ref PRO 30/82/1.
21. R.B. Russell, *Robert Aickman – An Attempted Biography* (Tartarus Press, Leyburn 2022) p.8.
22. Rolt to Aickman, 25 November 1945, National Archives, Aickman papers, ref PRO 30/82/1.
23. Rolt, L.T.C., RfD p.200.
24. Rolt to Aickman, 22 June 1946, National Archives, Aickman papers, ref PRO 30/82/2.
25. Hadfield to Ray Aickman, 14 May 1946, Inland Waterways Museum archives, Ellesmere Port, ref Hadfield 10/16b.

26. Rolt to Aickman, 16 July 1946, National Archives, Aickman papers ref PRO 30/82/2.
27. Aickman to Rolt, 23 July 1946, National Archives, Aickman papers, ref PRO 30/82/2.
28. Elizabeth Jane Howard, *Slipstream – a Memoir* (Macmillan, London, 2002) pp.196–99.
29. Rolt to Hadfield, 24 October 1946, Inland Waterways Museum archives, Ellesmere Port, ref Hadfield 10/16b.
30. Rolt to Aickman, 1 September 1946. National Archives, Aickman papers, ref PRO 30/82/2.
31. LWC p.89.
32. Rolt to Ray Aickman, 19 October 1946, National Archives, Aickman papers, ref PRO 30/82/2; Rolt to Robert Aickman, 22 October 1946, National Archives, Aickman papers, ref PRO 30/82/2.
33. Angela Rolt to Aickman, 19 November 1946, National Archives, Aickman papers, ref PRO 30/82/2.
34. Messrs Richard Marsh Ltd to BBC, 7 November 1946, National Archives, Aickman papers, ref PRO 30/82/2.
35. David Bolton, *Race Against Time – how Britain's Waterways Were Saved* (London: Methuen, 1990) gives the Memorandum at Appendix A, pp.257–8.
36. Ibid, pp.29–30.
37. *Daily Herald*, 6 March 1947.
38. Inland Waterways Association: correspondence and literature on the aims and works of the association, National Archives, Kew, Ref MT 52/213, 'Notes of meeting in the minister's room at 3pm on Wednesday, 6 March, 1947'.
39. Aickman to Rolt, 18 Jan 1947, National Archives, Aickman papers, ref PRO 30/82/3.
40. LWC pp.98 and 103.
41. *Birmingham Daily Gazette*, 19 May 1947.
42. Rolt to Aickman, 6 January 1947, National Archives, Aickman papers, ref PRO 30/82/3.
43. *Birmingham Post*, 19 March 1947.
44. Frustratingly, Hansard does not appear to provide any record of this exchange.
45. Rolt to Aickman, 22 March 1947, National Archives, Aickman papers, ref PRO 30/82/3.
46. Rolt to Aickman, 12 May 1947, National Archives, Aickman papers, ref PRO 30/82/3.
47. LWC p.98.
48. LWC p.100.
49. Ibid.

50. LWC p.101.
51. TRUU p.61.
52. '*Cressy* log', 4 July 1947.
53. CY p.78.
54. Rolt to Aickman, 4 July 1947, National Archives, Aickman papers, ref PRO 30/82/3.
55. Rolt to Aickman, 10 July 1947, National Archives, Aickman papers, ref PRO 30/82/3.
56. Rolt to Hadfield, 9 August 1947, Archive of National Waterways Museum, Ellesmere Port, ref Hadfield 10/16b.
57. LWC p.71.
58. L.T.C Rolt, *High Horse Riderless* (London: George Allen & Unwin, 1947) p.65.
59. Ibid, p.171.
60. Rolt to Aickman 17 July 1947, National Archives, Aickman papers, ref PRO 30/82/3. Willans' nickname 'the Commander' appears to date from his wartime service with the Royal Naval Reserves.
61. Rolt to Aickman, 10 July 1947, National Archives, Aickman papers, ref PRO 30/82/3.
62. LWC pp.114–115
63. LWC p.115.
64. LWC pp.115–7; '*Cressy* log,' 17 – 20 August 1947.

Chapter Nine

1. Rolt to Aickman, 14 February 1947, National Archives, Aickman Papers Ref PRO 30/82/3.
2. TRRU, p.88.
3. Raymond Russell, *Robert Aickman: An Attempted Biography* (Tartarus Press, Leyburn, 2022) p.106.
4. Rolt to Hadfield, 21 December 1947, National Waterways Museum Archives, Ellesmere Port, ref Hadfield 10 / 16b.
5. Rolt to Aickman, 8 June 1948, National Archives, Aickman Papers Ref PRO/30/ 82/4.
6. Rolt to Aickman, 20 October 1947 National Archives, Aickman Papers, PRO/30/ 82/3.
7. Rolt to Ray Aickman, 22 October 1947, National Archives, Aickman Papers Ref PRO 30/82/3.
8. Bolton (1990), pp.41–2
9. Rolt to Hadfield, 20 March 1950, National Waterways Museum Archives, Ellesmere Port, ref Hadfield 10 / 16b.

10. The locks built to accommodate the northern keels were often no more than sixty feet in length. (LWC p.134.)
11. LWC p.135.
12. Rolt to Aickman, 10 August 1948, National Archives, Aickman Papers Ref PRO/30/ 82/4.
13. LWC pp.128–34. The Severn Railway Bridge was demolished in the 1960s.
14. TRRU pp.139–43 and R.B. Russell, *Robert Aickman: An Attempted Biography.* (Leyburn: Tartarus Press, 2022) pp.129–30.
15. James Sutherland, 'Tom Rolt: A Journey through the Standedge Tunnel,' Remembering Rolt Symposium 2010, in *Industrial Archaeology Review*, 33: 1, 2011, pp.8–10, quoting from p.9; LWC p.137.
16. TRRU p.145.
17. LWC pp.140–41.
18. Elizabeth Jane Howard, *Slipstream – A Memoir* (Macmillan, 2002) pp.204–5.
19. LWC p.142.
20. www.gracesguide.co.uk/Basingstoke_Canal and www.getsurrey.co.uk/news/local-news/walton-weybridge-local-history-society-4836778.
21. Bolton, (1990), p.52.
22. *Daily News* (London) 2 March 1949.
23. Bolton, (1990), p.55.
24. For example, *Dundee Evening Telegraph*, 1 March 1949; *Leicester Daily Mercury*, 2 March 1949; and *Truth*, 11 March 1949.
25. *Daily News* (London) 2 March 1949.
26. Rolt to Aickman, 9 March 1949, National Archives, Aickman Papers, PRO/30/ 82/5.
27. Rolt to Aickman, 26 March 1949, National Archives, Aickman Papers, PRO/30/ 82/5.
28. TRRU, p.157.
29. Kyrle Willans, 'Future of an Old Canal'. First published in *Country Life*, 24 December 1948; reprinted in *Basingstoke Canal News*, Autumn 2013, pp.18–20.
30. R.B. Russell, *Robert Aickman: An Attempted Biography* (Tartarus Press, Leyburn, 2022) p.137; Rolt to Aickman, 26 March 1949, National Archives, Aickman Papers, PRO/30/ 82/5; Bolton (1990), p.56.
31. Aickman to Rolt, undated but evidently July 1949, National Archives, Aickman Papers, PRO/30/ 82/5.
32. Rolt to Aickman, 16 August 1949, National Archives, Aickman Papers, PRO/30/ 82/5.
33. Rolt to Aickman, 20 August 1949, National Archives, Aickman Papers, PRO/30/ 82/5.

Chapter Ten

1. Rolt to Aickman, 28 April, 1949. National Archives, Aickman Papers, PRO/30/ 82/5.
2. Rolt to Aickman, 2 June 1949. National Archives, Aickman Papers, PRO/30/ 82/5.
3. '*Cressy* log', 29 May – 5 June 1949.
4. '*Cressy* log', 13 June 1949. Tom's Plaid Cymru membership card is in IGMT's Rolt Collection, ref ROLT 14/1/01.
5. LWC p. 151; '*Cressy* log', 13 June 1949.
6. LWC p148; '*Cressy* log', 9 June 1949.
7. '*Cressy* log', 15 June 1949.
8. Rolt to Aickman, 18 June 1949, National Archives, Aickman Papers, PRO/30/ 82/5.
9. LWC, illustration 11.
10. '*Cressy* log', 26 July 1949.
11. LWC p.182.
12. Hadfield to Rolt, 9 July 1949, Letter and enclosed notes on ms of *Inland Waterways of England,* IGMT Rolt Collection ref ROLT 01/5B/01 (25)
13. Rolt to Aickman, 2 July 1949, National Archives, Aickman Papers, PRO/30/ 82/5; LWF, p.11.
14. LWF p.10
15. Rolt to Aickman, 7 August 1949, National Archives, Aickman Papers, PRO/30/ 82/5.
16. LTC Rolt, *Railway Adventure* pp.45–6.
17. Rolt to Aickman, 7 September 1949, National Archives, Aickman Papers, PRO/30/ 82/5.
18. LWF pp.11–12.
19. '*Cressy* log', 21 September and 3 October 1949. Geoffrey had married Maria Behr-Loewenstein of Vienna in November 1944 (*Chester Chronicle*, 18 November 1944).
20. *Chester Chronicle*, 22 October 1949.
21. Rolt to Aickman, 20 October 1949, National Archives, Aickman Papers, PRO/30/ 82/5.
22. Ibid.
23. Rolt to Aickman, 22 October 1949, National Archives, Aickman Papers, PRO/30/ 82/5.
24. LWC p168.
25. Rolt to Aickman, 25 November 1949, National Archives, Aickman Papers, PRO/30/ 82/5.
26. Rolt to Aickman, 1 December 1949. National Archives, Aickman Papers, PRO/30/ 82/5.

27. Rolt to Aickman, 8 December 1949. National Archives, Aickman Papers, PRO/30/ 82/5.
28. Ibid.
29. LWC p.169.
30. Rolt to Aickman, 11 November 1949 and 20 November 1949, National Archives, Aickman Papers, PRO/30/ 82/5.
31. Rolt to Aickman, 6 February 1950, National Archives, Aickman Papers, PRO/30/ 82/6.
32. Aickman to Rolt, 7 February 1950, National Archives, Aickman Papers, PRO/30/ 82/6.
33. Bolton (1990) p.73.
34. Ray Aickman to Rolt, 7 February 1950, National Archives, Aickman Papers ref PRO 30/82/6.
35. Rolt to Ray Aickman, 9 February 1950, National Archives, Aickman Papers ref PRO 30/82/6.
36. LWC p170.
37. *London Gazette*, 1 January 1947 – Supplement, p.14; and Warwick University Library – Guide to Archives of the Transport and General Workers' Union. (https://warwick.ac.uk/services/library/mrc/research_guides/tgwu/.)
38. Rolt to Aickman, 12 September 1949, National Archives, Aickman Papers, PRO/30/ 82/5.
39. LWF pp.26–27.
40. The *Nottingham Evening Post* 18 August 1950 lists Vigo's play; the *Loughborough Echo* 18 August 1950 gives Sutro's piece. Perhaps performance of the two works alternated.
41. LWC, p.173.
42. Artemis Cooper, *Elizabeth Jane Howard – A Dangerous Innocence* (John Murray; London, 2016) p.96.
43. Elizabeth Jane Howard, *Slipstream – A Memoir* (Pan: London, 2002) p.212.
44. '*Cressy* log', entries for 8, 13 and 14 May 1950.
45. L.T.C. Rolt, *The Thames from Mouth to Source* (London: Batsford, 1951) pp.37, 55 and 63.
46. LWC p.174.
47. LWC p.177.
48. Rolt to Aickman, 20 June 1950. (National Archives, Aickman Papers, PRO/30/ 82/6 and Bolton (1990) p.78.
49. Russell (2022) p.153. Given source Angela Rolt to Ray Aickman, 28 June 1950.
50. Rolt to Ray Aickman, 28 June 1950.
51. Philip Unwin to Rolt, 25 July 1950, Reading University Library Special Collections Archive of British Printing and Publishing, Allen & Unwin Letters.

Notes

52. Rolt to Aickman, 19 July 1950, National Archives, Aickman Papers, PRO/30/ 82/6.
53. Bolton (1990) p.81.
54. LWC p.181.
55. TRRU p.199.
56. *Leicester Daily Mercury*, 15 August 1950.
57. TRRU p.199.
58. *Leicester Daily Mercury*, 24 August 1950.
59. LWC p.181.
60. Rolt to Unwin, 29 August 1950, Reading University Special Collections, Archive of British Printing and Publishing Archive, Allen & Unwin letters.
61. LWC p.98, footnote and www.broadlandmemories.co.uk/wherrydragonlogs.htm.
62. LWC pp.182–3.
63. Philip Unwin to Rolt, 23 November 1950. Reading University Special Collections, Archive of British Printing and Publishing, Allen & Unwin letters. Unwin asks whether, for future correspondence with Tom, he should use the Banbury High Street address or 'the Cheltenham one', that is, Stanley Pontlarge. Tom's reply does not appear to survive.
64. Quoted by Bolton, (1990) p.86.
65. LWC p.186; Russell (2022) p.161.
66. LWC p.185.
67. Russell (2022) p.161.
68. Russell (2022) p.164; Bolton (1990) p.88.
69. Undated draft of a letter from Sonia Rolt to Angela Rolt, probably written in the mid-1970s, about the time of *Landscape with Canals'* publication. IGMT Rolt Collection, ref ROLT 09/2/04. .
70. Russell (2022) p.164; Bolton (1990) pp.89–90.
71. Bolton (1990) p.91.
72. CY p.99.
73. LWC p.183.
74. Rolt to Aickman, 12 February 1946, National Archives, Aickman Papers, PRO/30/ 82/2.
75. LWM p.227.
76. Rolt to Aickman, 10 December 1946, National Archives, Aickman Papers, PRO/30/ 82/2.
77. Artemis Cooper, *Elizabeth Jane Howard – A Dangerous Innocence* (John Murray, 2016) p.70.
78. Angela Rolt to Sonia Rolt, 6 March 1976, IGMT Rolt Collection, ref ROLT 09/2/04.
79. CY p.87.
80. LWC p.183–84.

81. www.encyclopedia-titanica.org/community/threads/helen-russell-cooke.8577/.
82. '*Cressy* Log', final page. (Undated.)
83. Harry Arnold, 'Cressy's Tale' in *NarrowBoat*, Summer 2010, pp.13–15, at p.15.
84. LWC p.187. Rolt's letters to Unwin of 16, 21 and 27 February, 17 March, and 19 and 24 April 1951 all bear Trinder's address 'C/o 84 High Street, Banbury, Oxon. Reading University Special Collections, Archive of British Printing and Publishing, Allen & Unwin letters. The premises now carries a plaque commemorating the two men's contribution to railway preservation. See https://banburyrotaryclub.org.uk/2015/past-president-bill-trinder/.

Chapter Eleven

1. Rolt to Unwin, 2 June 1951, Reading University Special Collections, Archive of British Printing and Publishing, Allen & Unwin letters.
2. Unwin to Rolt, 21 August 1950, Reading University Special Collections, Archive of British Printing and Publishing, Allen & Unwin letters.
3. Rolt, *Railway Adventure*, p.46.
4. Rolt, *Railway Adventure* p.8; L.T.C Rolt 'The Talyllyn Railway' in *Transactions of the Newcomen Society* vol. 33, 1960 – Issue 1, pp.17–29 at p.24–5. Given source *Lowca Engineering Company Catalogue*, nd.
5. www.steamlocomotive.com/appliances/valvegear.
6. L.T.C. Rolt, 'The Talyllyn Railway' in *Transactions of the Newcomen Society* vol. 33, 1960 – Issue 1, p.23.
7. J.B. Snell, 'The Bryneglwys slate quarry' in *Talyllyn Century*, (1965), from *Talyllyn Adventure* (Newton Abbot: David & Charles, 1971) pp.180–192 at p.181; and https://www.llechi.cymru/slateareas/bryneglwys.
8. Rolt, *Railway Adventure* p.23.
9. Snell (1965) p.189.
10. Graham Jones, 'To Hold the old Flag – Sir Henry Haydn Jones, 1863 – 1950', in *Journal of Liberal History* (Spring 2006) vol. 50, pp.20–25 and Wikipedia entry for Sir Henry Haydn Jones.
11. Snell (1965) pp.190-91; Jones (2006) p.23.
12. LWF pp.3–5.
13. Rolt, *Railway Adventure* p.43.
14. Rolt, *Railway Adventure* p.46–7.
15. LWF p.11.
16. Sir Henry Haydn Jones to Trinder, 25 November 1949, IGMT Rolt Collection ref R 2/3B/02/03.

Notes

17. Trinder to Sir Henry Haydn Jones, 9 December 1949, IGMT Rolt Collection, ref R 2/3B/02/03.
18. Rolt, *Railway Adventure* pp.41–2.
19. Trinder to Sir Henry Haydn Jones, 9 December 1949, IGMT Rolt Collection, ref ROLT 02/3B/02/03.
20. Trinder to Sir Henry Haydn Jones, 22 March 1950, IGMT Rolt Collection, ref ROLT 02/3B/02/03
21. *Liverpool Daily Post* 3 July 1950.
22. LWF p.12.
23. L.T.C. Rolt, letter to various correspondents, 20 September 1950 quoted in full in LWF, p.14.
24. 'Minutes of the Meeting convened to consider the future of the Tal-y-Llyn Railway, at the Imperial Hotel, Temple Street, Birmingham on 11 October, 1950'. Reading University Special Collections, Archive of British Printing and Publishing, Allen & Unwin letters.
25. P.B. Whitehouse, 'Society and Company, 1950-1965' in *Talyllyn Century* (1965), reprinted in *Talyllyn Adventure* (Newton Abbot: David and Charles, 1971), pp.214–26 at pp.214–15.
26. *Railway Magazine*, 19 December 2019.
27. Minutes of the Meeting [...] on 11 October 1950. (See note 24.)
28. Obituary for Barbara Curwen, *Wiltshire Gazette and Herald,* 23 August 2007.
29. James I.C. Boyd, 'The Talyllyn Railway 1865 – 1950' in *Talyllyn Adventure*, pp.193–213, at p.211.
30. Minutes of the Meeting [...] on 11 October 1950 (See note 24.)
31. Rolt, *Railway Adventure*, p.56.
32. J.L.H. Bate, 'Engineering Progress, 1951-1965' in *Talyllyn Adventure*, pp.227–43, at pp.228–31.
33. No-one foresaw it at the time, but in the second half of the 1950s funds would be found for *Talyllyn*'s restoration and the locomotive returned to service on 14 June 1958. L.T.C Rolt, 'Motive Power,' *Talyllyn Adventure*, pp.244–56, at p.248.
34. Rolt, *Railway Adventure*, p.59.
35. Rolt, *Railway Adventure*, p.61.
36. Rolt to Unwin, 16 February 1951, Reading University Special Collections, Archive of British Printing and Publishing, Allen & Unwin letters relating to L.T.C. Rolt, 1951.
37. Unwin to Rolt, 21 and 27 February 1951, Reading University Special Collections, Archive of British Printing and Publishing, Allen & Unwin letters relating to L.T.C. Rolt, 1951.
38. Rolt, *Railway Adventure*, p.62; and LWF p.17; and Rolt to Unwin, 17 March 1951, Reading University Special Collections, Archive of

British Printing and Publishing, Allen & Unwin letters relating to L.T.C. Rolt, 1951.
39. Rolt, *Railway Adventure*, p.63.
40. Rolt, *Railway Adventure*, p.124.
41. W. Emley of Johnson, Jecks and Landon, Solicitors, 29 March and 6 April 1951, IGMT Rolt Collection, ref ROLT 02/3B/02/02 (20)
42. Rolt, *Railway Adventure*, p.64. For further commentary on the circumstances in which Snell took over Dai Jones's job, see David Potter, *The Talyllyn Railway* (Newton Abbot: David St John Thomas, 1990) pp.120-22.
43. J.B. Snell, 'Memories of Tom Rolt' in *L.T.C. Rolt – A Bibliography*, compiled by Ian Rogerson, Gordon Maxim and Mark Baldwin, edited and revised by Tim Rolt and Andrew Thynne, (REM: Bath, 2012) pp.10–13.
44. LWF p.19.
45. Robert Humm, 'P.B. Whitehouse' in *Railway Magazine*, 19 December, 2019 (www.railwaymagazine.co.uk/13361/from-the-archive-p-b-whitehouse/).
46. www.fairbournerailway.com/history.html.
47. www.therailwayhub.co.uk/38473/the-last-haydn-jones-train-commemoration/.
48. LWF p.19.
49. Rolt, *Railway Adventure*, p.132.
50. *Talyllyn News*, No. 4, June 1954, p.6.
51. Rolt to Unwin, 4 July 1951, Reading University Special Collections, Archive of British Printing and Publishing, Allen & Unwin papers.
52. LWC p.185.
53. LWF pp 30–1. For a rather different version of events, see Tim Coghlan, 'Sonia Rolt OBE, 1919-2014', in *IWA Waterways Magazine*, Spring 2015, pp.21–7, at p.25.
54. Snell (2012) pp.10–13.
55. LWF pp.41 and 57; *MotorSport Magazine*, August 1940, under 'Club News'.
56. Snell (2012) p.11.
57. LWF pp. 30–1.

Chapter Twelve

1. LWF pp.23, 31–5.
2. LWF pp.35–6.
3. Rolt to Philip Unwin, 6 July 1951, Reading University Special Collections, Archive of British Printing and Publishing, Allen & Unwin papers.

Notes

4. Unwin to Rolt, 31 January 1952, Reading University Special Collections, Archive of British Printing and Publishing, Allen & Unwin papers.
5. Rolt to Unwin, 14 March 1952, Reading University Special Collections, Archive of British Printing and Publishing, Allen & Unwin papers.
6. LWF pp.93–101 gives a full account of the trip. Tom and Sonia did not, in fact, marry until 1969, although to quell the disapproval of certain TRPS members who had raised objections to their co-volunteering for the 1952 season Sonia changed her surname to 'Rolt' by deed poll. Apparently one or other of them had difficulty in obtaining a divorce. When their wedding eventually took place, it was in a London Register Office and was a legal formality with the minimum required number of witnesses present. LWF, pp.37 and 93; C.S Clutton to Sonia Rolt, 19 February 1974, IGMT Rolt Collection ref R 1/5D/01; BMD and author's exchange with Tim Rolt, March 2023.
7. https://en.wikipedia.org/wiki/Tralee_and_Dingle_Light_Railway.
8. LWF p.97. For the Curraduff (Camp) and Lispole derailments see 'Traces of the Dingle train' on www.narrowgauge.nl/site/english/tdsurvey.htm.
9. LoC pp.177–82
10. LWF p.98.
11. LWF p.100.
12. LoC p.177.
13. *Birmingham Daily Gazette*, 30 December 1952; *Crewe Chronicle*, 29 November 1952; *Western Mail*, 19 November 1952.
14. LWF p.126.
15. LWF pp.37–8.
16. Reverend W. Awdry, 'Talyllyn Impressions 1952', short unpublished memoir, supplied by Mr Keith Theobald of the Narrow Gauge Railway Museum, Towyn, and used here by permission of Christopher Awdry.
17. Brian Sibley, *The Thomas the Tank Engine Man – the Life of Reverend W. Awdry* (Lion Hudson, Oxford, 2015) p.193.
18. Ibid, p.189.
19. Ibid, pp.110–11. For a discussion of Tom's early railway stories, see Victoria Owens '"The Railway": an unknown early work by LTC Rolt', *Journal of the Railway and Canal Historical Society*, Vol 41, Part 1, No 246, March 2023, pp 2-13.
20. Ibid, pp.190–91.
21. Reverend W. Awdry, *Four Little Engines* (Leicester: Edmund Ward Ltd, 1955) pp.34–48.
22. LWF pp.42–43. Tom to Bill Harvey on 30 December 1952, IGMT Rolt Collection, ref ROLT 02/3B/02/06. .
23. LWF p.126.
24. Rolt, *Railway Adventure*, pp.154–57.

25. Canon Roger Lloyd, 'Rescuing a Railway' in *Truth*, 8 January 1954, p.24.
26. LWF pp.44–6. Addresses given in TR-related correspondence between Tom and 'Bill' (Harvey) 30 December 1952 and 20 July 1953, IGMT Rolt Collection ref R 2/3B/02/06.
27. Rolt to Bill Harvey, 7 May 1953, IGMT Rolt Collection ref ROLT 02/3B/02/06..
28. *Talyllyn News*, No.1, September 1953, p.15.
29. Rolt to Bill Harvey, 20 July 1953, IGMT Rolt Collection ref R 2/3B/02/06..
30. C. Uren, 'No Room for Complacency', in *Talyllyn News*, No.2, November 1953, pp.2–4.
31. 'Allan' [Garraway] to Rolt, 4 September 1953, IGMT Rolt Collection, ref ROLT/02/3B/02/03/20.
32. *Talyllyn News*, March 1958, p.7.
33. LWF p.46.
34. The date-range tallies with Tom's recollection that work began on his mother's dower house 'in the late autumn of 1953' leading to her temporarily moving in 'with a friend in the neighbouring village' (LWF p.48).
35. LWF p.48.
36. Ibid, pp.50–1.
37. L.T.C. Rolt, *Winterstoke* (Constable: London, 1954) pp.232–3.
38. *Tewkesbury Register* 16 October 1954; *Birmingham Daily Post*, 13 July and 30 November 1954.
39. LWF p.128.
40. https://faberfinds.wordpress.com/about/.
41. LWF pp.128–9.
42. LWF pp. 53–4.
43. *Sunderland Daily Echo and Shipping Gazette*, 13 August 1954.
44. *Lancaster Guardian*, 3 September 1954.
45. LWF pp.68–9.
46. *Coventry Evening Telegraph*, 2 September 1954; *Birmingham Daily Post*, 3 September 1954, *et al*. The appearance of Henry Cornelius' film *Genevieve* the previous year which concerned events connected with the London-Brighton run may have boosted the enthusiasm which greeted the rally.
47. *Daily News* (London), 6 September 1954.
48. *Birmingham Weekly Post*, 10 September 1954 and LWF p.63.
49. LWF p.64.
50. LWF p.66.
51. LWF pp.70–1.
52. *Daily News* (London), 8 September 1954.
53. LWF pp.74–5.

54. LWF p.76.
55. *Shields Daily News*, et al, 11 September 1954.
56. LWF p.77. The Americans got their revenge at a follow-up rally in the US two years later, where they trounced their British rivals.

Chapter Thirteen

1. Rolt to Aickman, 10 July 1947, National Archives, Aickman Papers PRO 30/82/3.
2. Sonia thought that Charles Hadfield's friend Frank Eyre had been responsible for introducing Tom to Hough. 'Tom Rolt – A Life in Books', *Waterways World*, January 2010, pp.57–62 at p.59.
3. https://en.wikipedia.org/wiki/Harrow_and_Wealdstone_rail_crash.
4. *Dundee Courier*, 10 October 1952.
5. Rolt, *Railway Adventure*, p.151.
6. LWF p.129; and www.steamindex.com/people/inspoff.htm which includes entries for Col. McMullen and Lt-Col. Wilson.
7. LWF p.20.
8. LWF p.24.
9. Jack Simmons & Gordon Biddle (eds.), *Oxford Companion to British Railway History* (Oxford: Oxford University Press, 1997) entry for 'Trade, Board of'.
10. RfD p16 and www.orr.gov.uk/search-news/story-hmri-video-history-her-majestys-railway-inspectorate.
11. *The Railway Yearbook*, published by the Railway Publishing Company, seems to have been relatively short-lived but flourished in the 1910s –1920s. Tom evidently had a copy (year unspecified) in his possession (LWF p.130).
12. In the preface to *Red for Danger*'s first edition, Tom explicitly confines his coverage to accidents 'from 1840 to 1940' on the ground that he thought it 'inappropriate' to write about more recent disasters. Later editions of the book address the 1952 Harrow and Wealdstone crash. (My thanks to John Freeman for confirming this point.)
13. Shakespeare, *Henry IV pt 1*, Act II, sc.3.
14. RfD p.125 and www.railwaysarchive.co.uk/.
15. RfD p.18.
16. RfD p.40.
17. RfD pp.187–93, at p.93.
18. LoC p.56
19. RfD p.150.
20. RfD pp.147–153.

21. RfD p.154.
22. *LoC* p.74.
23. Ibid.
24. RfD, p.160.
25. LWF p.130 and RfD Preface *passim*.
26. Entries for *Red for Danger* in *L.T.C. Rolt – A Bibliography,* compiled by Ian Rogerson, Gordon Maxim and Mark Baldwin, edited and revised by Tim Rolt and Andrew Thynne, (REM: Bath, 2012).
27. www.railpro.co.uk/railpro-magazine/may-2022/remembering-to-remember.
28. http://publictransportexperience.blogspot.com/2012/09/l-t-c-rolt-and-teaching-career.htm.
29. *Guardian*, 10 March 1989, Obituary for Frank Rudman.
30. Rudman to Rolt, 30 August 1953, IGMT Rolt Collection, ref ROLT 01/5B/05.
31. LWF pp.133–5.
32. James Dugan, *The Great Iron Ship* (Hamish Hamilton: London, 1953); LWF p.142–33.
33. Henri Delgove to Rolt, nd. but prob. 1955, IGMT Rolt Collection ref ROLT 04/50/04. .
34. *The Independent*, 16 July 1997, obituary for George Hardinge (Lord Hardinge of Penshurst).
35. LWF p.143.
36. IK Brunel's granddaughter, Celia James, married an Eton College housemaster named Saxton Noble. Sir Humphrey Noble was their son and IK Brunel's great grandson. Rather confusingly Sir Humphrey married another Celia – Celia Weigall – so his mother and wife shared the same first name. See Isambard Kingdom Brunel: Family History | Tracing Ancestors In The UK (tracingancestors-uk.com) and https://www.geni.com/people/Celia-Noble/6000000130457858089.
37. Tom describes his visit in LWF pp.136–40.
38. Rolt, *Isambard Kingdom Brunel* p.402; Sir Humphrey Noble to LTCR, 1 December 1955, IGMT Rolt Collection ref R 4/50/04.
39. Rolt, *Isambard Kingdom Brunel*, p.403.
40. Sir Humphrey Noble, letter to LTCR, 1 December 1955, IGMT Rolt Collection ref ROLT 04/50/40.
41. L.T.C. Rolt, *George and Robert Stephenson* (1960) p.59. The mistake may arise from a confused recollection that Brindley married Anne, sister of his assistant Hugh Henshall. In fact, the couple's older daughter Ann never married while the younger daughter Susannah became the wife of John Bettington of Bristol. See Victoria Owens, *James Brindley and the Duke of Bridgewater – Canal Visionaries* (Amberley Books, 2015), pp.125–7.

Notes

42. LWF p.139.
43. https://collection.sciencemuseumgroup.org.uk/people/ap24885/russell-john-scott.
44. LWF p.133.
45. Adrian Vaughan, *Isambard Kingdom Brunel – Engineering Knight Errant* (London: John Murray, 1991) Preface, *citing* Rolt, *Isambard Kingdom Brunel* (1955).
46. Vaughan (1991) page xi.
47. Angus Buchanan, *Brunel* (Hambledon and London, London, 2002) p.211.
48. Rolt, 'In the footsteps of Samuel Smiles' *The Author*, (journal of the Society of Authors) Winter 1970, pp.151–55.
49. LWF, p.146.
50. *Newcastle Journal*, 21 May 1965; incidentally, the newspaper mistakenly gives the surname 'Rolt' as 'Bolt' but its description of Tom as 'having published more than a score a of successful books on subjects which range from philosophy to motoring' leaves no room for doubt as to whom they mean.
51. Simon Garfield, *The Last Journey of William Huskisson* (Faber and Faber, London, 2002.)
52. Julian Glover, *Man of Iron: Thomas Telford and the building of Britain* (Bloomsbury, London, 2017.)

Chapter Fourteen

1. *Cornish Guardian*, 4 August, 1960.
2. Rolt, L.T.C *Great Engineers* (George Bell & sons: London, 1962; repr.1966), Foreword, page v.
3. *Birmingham Daily Post*, 6 February 1962; *Daily Herald*, 9 February 1962.
4. L.T.C. Rolt, *James Watt* (Batsford, 1962) p.60.
5. LWF p.158.
6. Samuel Smiles, *James Brindley and the Early Engineers* (John Murray, 1864) pp.19 and 114–6.
7. Rolt, 'Samuel Williams & Sons Ltd; 1855-1955' from *A Company's Story in its Setting – Samuel Williams & Sons Ltd 1855-1995* (London: Newman Neame Ltd, 1955) p.50. The book contains five chapters by different authors.
8. Ibid, p.78.
9. LWF p.160–61.
10. Pocket appointment diary, 1958, IGMT Rolt Collection, ref ROLT 01/6B/04, , entries made for 27 January -14 February 1958.
11. LWF p.166; for further information – specifically about the Baldwin locomotives – see also www.internationalsteam.co.uk/trains/dominicanrep05.htm.

12. https://en.wikipedia.org/wiki/Santo_Domingo.
13. LWF pp.160–68.
14. LWF pp.168–75.
15. LWF pp.174–5.
16. LWF p.173. A painting of the ship is in the collection of Whitehaven's Beacon Museum. https://artuk.org/discover/artworks/emily-burnyeat-143341.
17. Peter Roberts, *L.T.C. Rolt – A Collector's Bibliography*, Private Publication, 2021; ISBN 978 1 3999 -1237 -2.
18. LWF p.187.
19. L.T.C. Rolt, *A Hunslet Hundred – One Hundred Years of Locomotive Building by the Hunslet Engine Company*, (David & Charles (Dawlish) and MacDonald (London) 1964); *Waterloo Ironworks – A History of Taskers of Andover, 1809-1968*, (David & Charles (Newton Abbot) 1969).
20. E. Temple Thurston, *The 'Flower of Gloster'* 1st published 1911, reissued with an introduction by L.T.C. Rolt, (David and Charles: Newton Abbot, 1968)
21. LWF pp.187–8.
22. Ed Howker, 'Asbestos – the lies that killed', *New Statesman*, 28 August, 2008.
23. Author's conversation with Tim Rolt; Peter Roberts, *L.T.C. Rolt – A Collector's Bibliography* (Private Publication, 2021) p.52.
24. *Turner and Newall Limited: The First Fifty Years 1920–1970* (1970). Copyright in the printed book is assigned to Turner and Newall Limited; it makes no mention of the author.
25. LWF pp.149–52. Under the 1979 Public Lending Right Act, British authors have a legal right to receive payment for the free loan of their books from public libraries. https://www.bl.uk/plr/about.
26. LWF p.122.
27. LWF pp.216–17.
28. Angus Buchanan, 'The origins and early days of the AIA,' *Industrial Archaeology News* 169, Summer 2014 pp.2–4, quoting from p.2.
29. 'John Betjeman, the Euston Arch and the Fight to Save London's Industrial Heritage', lecture by Dr Ruth Adam, 30 April 2017. Text available at: https://chmcc.hypotheses.org/2663.
30. Angus Buchanan, 'The origins and early days of the AIA', *Industrial Archaeology News* 169, Summer 2014, pp.2–4 and 'The Centre for the Study of the History of Technology, Bath University of Technology, England', *Technology and Culture*, 1968, pp.430–35.
31. *Bristol Evening Post*, 6 July 1973; and Buchanan (2014) p.3
32. LWF p.212.

33. Anthony Bird, *Roads and Vehicles* (1969); W.A. Campbell, *The Chemical Industry* (1971); and Kenneth Hudson, *Building Materials* (1972) all published within the Longman Industrial Archaeology series.
34. L.T.C. Rolt, *The London-Birmingham Motorway* (Ministry of Transport and Civil Aviation, London, c.1960); *The Tyne Tunnel* (Tyne Tunnel Joint Committee, Newcastle, 1967); and *Mersey Tunnel 2* (Mersey Tunnel Joint Committee, Liverpool, c.1971).
35. Tredgold conceived his statement for the newly founded Institution of Civil Engineers who required a concise summary of their work in order to obtain a royal charter. Rolt, *Navigable Waterways* (Longman: London, 1969), Introduction p.x.
36. Rolt, *Navigable Waterways* (Longman: London, 1969), Introduction p.x.
37. Rolt, *Victorian Engineering* (Penguin, 1970) p.281 quoting Francis Klingender, *Art and the Industrial Revolution* (1947), p.27.
38. Rolt, *Victorian Engineering*, p.283.
39. Rolt, 'Strange Vista', Red Book Manuscript, Part I, Chapter 7.

Chapter Fifteen

1. Tim Rolt, email to the author, 2 December 2022 and BBC Radio Four programme celebrating the life and legacy of L.T.C. Rolt broadcast on 8 November 2010, presented by Hermione Cockburn and produced by Julian May.
2. Rolt to Hadfield, 7 December 1954, Inland Waterways Museum Archives, Ellesmere Port, ref Hadfield 10/16b.
3. L.T.C. Rolt, 'In the Steps of Samuel Smiles,' *The Author*, vol. LXXXI, No. 4, Winter 1970, pp.151–55.
4. LWF pp. 153 and 155.
5. Correspondence between Rolt and Faber & Faber, IGMT Rolt Collection, ref R 3/1/04.
6. Hadfield to Rolt, 14 August 1969, IGMT Rolt Collection ref ROLT 03/1/08.
7. Hadfield to Rolt, 19 October 1970, IGMT Rolt Collection ref ROLT 03/1/08.
8. IGMT Rolt Collection ref ROLT 03/1/01.
9. Judith Niechcial, *A Particle of Clay – The Biography of Alec Skempton, Engineer* (Whittles Publishing: Latheronwheel, Caithness, 2002), p.166.
10. Ward Lock Letters, Platt to Rolt, 15 March 1973, et seq., IGMT Rolt Collection, ref ROLT 03/1/04.
11. Heinemann Educational Books correspondence, IGMT Rolt Collection, ref ROLT 03/1/04.

12. MacDonald & Co letters, IGMT Rolt Collection, ref ROLT 03/1/04. Ruth Thomson's book *The Stephensons* (London: Macdonald & Co., 1974) was one of MacDonald's 'Starters' series.
13. For an account of the consequences that the Canal du Midi held for British canal construction see Nicholas Hammond, 'Francis Egerton's Visit to the Canal du Midi in 1754 and its extraordinary aftermath', in *Railway and Canal Historical Society Journal*, vol. 38 Part 7, (March 2016), pp.440–49.
14. Rolt, engagement diary for 1971, IGMT Rolt Collection ref ROLT 01/6B/17 and *From Sea to Sea* (Seyssinet, France: Europmapping, 1994; First published Allen Lane, London, 1973) p.194.
15. Ibid, p.198.
16. Ibid, pp.176 and 179.
17. Page proofs for *From Sea to Sea,* IGMT Rolt Collection, ref R 4/10/04 -05.
18. Rolt, Engagement Diary for 1972, IGMT Rolt Collection, ref ROLT 01/6B/18.
19. *Coventry Evening Telegraph*, 21 April 1972.
20. There is a short biography of Fairgrieve on the 'Plarr's Lives of the Fellows' section of the Royal College of Surgeons website: www.rcseng.ac.uk (see Fairgrieve, John 1926–2014).
21. Rolt, Engagement Diary for 1972, IGMT Rolt Collection, ref ROLT 01/6B/18.
22. Harper to Rolt, 28 December 1972, Harper Letters in IGMT Rolt Collection, ref ROLT 01/5A/16. For Tom's account of their friendship, see LWF pp.222–24.
23. LWF p.237.
24. Foreword to LWF, page x.
25. Manuscript of *Landscape with Machines,* IGMT Rolt Collection, ref ROLT 04/60/03..
26. Goldsworthy Lowes Dickinson, *A Modern Symposium* (New York: Doubleday, Page & Co, 1905) pp.119-20, available online at https://archive.org.
27. Foreword to LWF, page x.
28. LWF p.235.
29. LWC p.182
30. LWF pp.225–29.
31. https://en.wikipedia.org/wiki/John_Smith_(Conservative_politician).
32. Bolton, 1990, p.250.
33. Ibid; and R.B. Russell, *Robert Aickman – An Attempted Biography* (Tartarus, 2022), p.271.
34. Russell (2022), p.272. (Russell's italics.)
35. The manuscript (typescript) of Landscape with Canals (pages not numbered) is in IGMT Rolt Collection ref ROLT 04/61/04.

Notes

36. LWC p.181.
37. Angela Rolt to Sonia Rolt, 18 November 1975, IGMT Rolt Collection ref ROLT 09/2/04.
38. Angela Rolt to Tom Rolt, 11 November 1973, IGMT Rolt Collection ref ROLT 01/5D/01(19). .
39. *Manchester Evening News*, 3 December 1973.
40. Barrie and Jenkins correspondence, specifically the letters of L.T.C Rolt and Aidan Chambers, esp. 28 Dec 1973 and 2 and 7 January 1974. IGMT Rolt Collection, ref ROLT 01/1/04. .
41. Information about Hugh Lamb's short story collections from https://hughlamb.com/bibliography/.
42. C. S. Clutton to Rolt, 9 March 1974, IGMT Rolt Collection, ref ROLT 01/5D/01.
43. C.S. Clutton to Sonia Rolt, 19 February 1974, IGMT Rolt Collection ref ROLT 01/5D/01.
44. Christian Smith to Sonia, 23 February 1974, IGMT Rolt Collection, ref R 01/5D/01.
45. Ecclesiasticus Ch 38, vv. 31–2.
46. Letters of condolence in IGMT Rolt Collection ref ROLT 01/5B/10.
47. The drawing – both original and copy – is in IGMT Rolt Collection, ref ROLT 01/5A/10 (Betjeman Correspondence).
48. Angela Rolt to Sonia Rolt, 4 June 1974, IGMT Rolt collection ref ROLT 01/5D/01(32)..
49. Robert Aickman to Sonia Rolt, 15 May 1974, IGMT Rolt collection ref ROLT 01/5D/01(282) .
50. Peter Carson of Penguin Books to Hilary Rubinstein – literary agent – of A.P. Watt & Son, copied to Sonia Rolt, 24 January 1975, IGMT Rolt collection ref ROLT 03/1/15.
51. Sonia Rolt to Messrs A.P. Watt & Son, 26 March 1975, IGMT Rolt Collection ref ROLT 03/1/15.
52. Julia Elton, Obituary for Sonia Rolt, OBE, FSA, *Newcomen Links*, 233, March 2015, pp.20–21.
53. The History Press sells a number of Rolt titles.
54. LWF p.246.
55. Guest's obituary of Rolt, *Report of the Royal Society of Literature*, 1974-5, pp.28–30.

Bibliography

Manuscripts

National Archives, Kew
Letters of Robert Aickman to L.T.C. Rolt: Correspondence and Papers of Robert Aickman. (Ref: PRO 30/82.)
University of Reading Special Collections, Archive of British Printing and Publishing Records of George Allen & Unwin. (Ref: AU.)

Waterways Archive, Ellesmere Port
Letters of L.T.C. Rolt to Charles Hadfield. (Ref: Hadfield 10/16b.)

Ironbridge Gorge Museums Trust Archives
Letters of L.T.C Rolt to Robert Aickman. (Ref: ROLT 06/IWA/28.)
Rolt, L.T.C, 'The *Cressy* Log': IGMT Rolt Collection. (Ref ROLT 01/6A/01.)
Rolt, L.T.C, 'The Railway, being a collection of Railway Stories, Verses and Anecdotes' unpublished manuscript: IIGMT Rolt Collection. (Ref ROLT 1/4A/03.)

* * * * * * *

Rolt, L.T.C., 'Strange Vista' Unpublished novel. The 'Red Book manuscript' c.1933–5, includes Prologue, 1760–1860; Part One, 1929; Part Two, 1954. The 'Green Book manuscript' [incomplete] c.1936–7, includes Prologue, 'Yesterday'; Part One, 'Today'. (Both mss in possession of the Rolt family.)

Printed Sources

Books by L.T.C. Rolt
(This list is by no means exhaustive)

From Sea to Sea (Euromapping, Seyssinet, France, 1994). (First published Allen Lane, London, 1973.)

Bibliography

George and Robert Stephenson – The Railway Revolution (Pelican, Harmondsworth, 1978). (First published Longman, London 1960.)
Great Engineers (George Bell & Sons, London, 1962).
Green and Silver (George Allen & Unwin, Ltd, London, 1949).
High Horse Riderless (George Allen & Unwin, Ltd, London, 1947).
A Hunslet Hundred (David & Charles, Dawlish, 1964).
Isambard Kingdom Brunel (First published Longmans Green, 1957; repr. Pelican, Harmondsworth, 1970).
The Inland Waterways of England (George Allen & Unwin, London, 1950).
James Watt (Batsford, London, 1962).
Landscape with Machines; an autobiography (First published London: Longmans, 1971; reissued Gloucester: Alan Sutton Publishing, 1984).
Landscape with Canals: being the second volume of *Landscape with Machines*, an autobiography (First published Allen Lane, London, 1977; reissued Gloucester: Alan Sutton Publishing, 1984).
Landscape with Figures; the final part of his autobiography (Gloucester: Alan Sutton, 1992).
Lines of Character (Constable, London, 1952).
Narrow Boat (Eyre & Spottiswoode, London, 1944).
Navigable Waterways (Longman, London, 1969).
Railway Adventure (First published Constable, 1953); reissued with *Talyllyn Century* (David and Charles, Newton Abbot, 1965) as combined volume entitled *Talyllyn Adventure*. (Newton Abbot: David & Charles, Newton Abbot, 1971).
Red for Danger (First published London: J. Lane, 1955; 4th edn. David & Charles, Newton Abbot, 1982).
'Samuel Williams & Sons Ltd; 1855–1955' from *A Company's Story in its Setting – Samuel Williams & Sons Ltd 1855–1995* (London: Newman Neame Ltd, 1955) pp.41-79.
Sleep No More (Stroud: The History Press, 2010). (First published London: Constable, 1948.)
The Thames from Mouth to Source (London: Batsford, 1951).
Victorian Engineering (Penguin, Harmondsworth, 1970).
Waterloo Ironworks – A History of Taskers of Andover, 1809-1968 (David & Charles, Newton Abbot, 1969).
Winterstoke (Constable: London, 1954).

Books by other authors
Aickman, Robert, *The River Runs Uphill* (Burton on Trent: J.M. Pearson, 1986).
Awdry, Rev. W, *Four Little Engines* (Edmund Ward, Leicester, 1955).
Bolton, David, *Race Against Time – How Britain's Waterways Were Saved* (London: Methuen, 1990).

Bomford, D.R., *Corn in England* (Private publication, 1927).

Buchanan, Angus, *Brunel* (Hambledon and London, London, 2002).

Cooper, Artemis, *Elizabeth Jane Howard – A Dangerous Innocence* (John Murray, London, 2016).

Dickinson, Goldsworthy Lowes, *A Modern Symposium* (New York: Doubleday, Page & Co, 1905).

Howard, Elizabeth Jane, *Slipstream – a Memoir* (Macmillan, London, 2002).

Jackson, Alan A., *The Railway Dictionary* (Alan Sutton Publishing Ltd., Stroud, 1992).

L.T.C. Rolt – A Bibliography, compiled by Ian Rogerson, Gordon Maxim and Mark Baldwin, edited and revised by Tim Rolt and Andrew Thynne (REM, Bath, 2012).

L.T.C. Rolt: A Collector's Bibliography, compiled by Peter R.G. Roberts (Orphans Press, Leominster, 2021).

Mackersey, Ian, *Tom Rolt and the Cressy Years* (M & M Baldwin, London, 1985).

Murray, Scott, *The Title – the Story of the First Division* (Bloomsbury, 2017) p.96.

Niechcial, Judith, *A Particle of Clay – The Biography of Alec Skempton, Engineer* (Whittles Publishing: Latheronwheel, Caithness, 2002).

Oxford Companion to British Railway History, edited by Jack Simmons & Gordon Biddle (Oxford: Oxford University Press, 1997).

Potter, David, *The Talyllyn Railway* (Newton Abbot: David St John Thomas, 1990).

Rolt, Col. Sir John, *On Moral Command* (London; W. Clowes, 1842). Available online via Google Books and Hathi Trust.

Ross of Bladensburg, Sir John Foster George, *A History of the Coldstream Guards* (A.D. Innes & Co., London 1896).

Russell, R.B., *Robert Aickman: An Attempted Biography* (Leyburn: Tartarus Press, 2022).

de Salis, H.R., *Bradshaw's Canals and Navigable Rivers of England and Wales – A Handbook of Navigation* (London: Henry Blacklock & Co Ltd., 1904).

Sherwen, Theo, *The Bomford Story* (Evesham: Bomford and Evershed Ltd., 1978).

Sibley, Brian, *The Thomas the Tank Engine Man – the life of Reverend W. Awdry* (Lion Hudson, Oxford, 2015).

Talyllyn Century (David & Charles, Newton Abbot, 1965) including essays by J.B. Snell, P.B. Whitehouse, L.T.C. Rolt et al. Reissued with *Railway Adventure* (1953) as a combined volume entitled *Talyllyn Adventure* (David & Charles, Newton Abbot, 1971).

Smiles, Samuel, *James Brindley and the Early Engineers* (John Murray, 1864).

Bibliography

The Future of the Waterways (pamphlet). No author given, but attributed to L.T.C. Rolt. (Inland Waterways Association, London, nd.).

Vaughan, Adrian, *Isambard Kingdom Brunel – Engineering Knight Errant* (London: John Murray, 1991).

Articles

Barnes, J.S, 'A Hanging Offence' *Times Literary Supplement*, 8 July 2022, p.26. (Review of Stephen Bates, *The Poisonous Solicitor*, London: Icon Books, 2022.)

Buchanan, Angus, 'The origins and early days of the AIA,' *Industrial Archaeology News* 169, Summer 2014. pp.2–4.

'The Centre for the Study of the History of Technology, Bath University of Technology, England' *Technology and Culture*, 1968, pp.430–435.

Coghlan, Tim, 'Sonia Rolt OBE, 1919-2014', *IWA Waterways Magazine*, Spring 2015, pp. 21–27.

Elton, Julia, Obituary for Sonia Rolt, OBE, FSA, *Newcomen Links*, 233, March 2015, pp.20–21.

Guest, John, Obituary for L.T.C. Rolt, *Report of the Royal Society of Literature*, 1974–5, pp.28–30.

Hammond, Nicholas, 'Francis Egerton's Visit to the Canal du Midi in 1754 and its extraordinary aftermath,' *Journal of the Railway & Canal Historical Society*, vol. 38 Part 7, (March 2016), pp.440–449.

Howker, Ed, 'Asbestos – the lies that killed', *New Statesman*, 28 August, 2008.

Jones, Graham, 'To Hold the old Flag – Sir Henry Haydn Jones, 1863 – 1950', *Journal of Liberal History* (Spring 2006) vol. 50, pp.20–25.

Minnis, John, 'Defining the Vintage Car: L.T.C. Rolt, the vintage sports car and the rise of motoring heritage,' *Industrial Archaeology Review*, 39, May 2017 pp.2–13.

Owens, Victoria, '"The Railway", an unknown early work by L.T.C. Rolt', *Journal of the Railway & Canal Historical Society*, vol. 41, Part 1, No. 246, March 2023, pp.2–13.

'L.T.C. Rolt's Stories of the Supernatural – *Sleep No More* (1948)' *Western Courier* 25, March 2020 (Newsletter of the Western Region of the Newcomen Society) pp.7–10.

'Remembering Rolt: A Symposium in Honour of the 100th Anniversary of L.T.C. Rolt's Birth – A Pioneer of Industrial Archaeology.' Papers by Angus Buchanan, Neil Cossons, Julia Elton, Keith Falconer, Richard Hope & James Sutherland (2011) *Industrial Archaeology Review*, 33:1, pp.3–17.

Rolt, L.T.C., 'Imagination and Dramatic Art', *Kingdom Come – the Magazine of Wartime Oxford*, (Summer 1941, pp.113–4).

'Ethics of the Quality Car', *MotorSport* (March 1945, pp.45–47).

'Planned tyranny', *Horizon* (January 1943, pp.1–5).

'The Talyllyn Railway', *Transactions of the Newcomen Society,* vol. 33, 1960 (1), pp.17–29.

'The History of the History of Engineering', *Transactions of the Newcomen Society*, vol. 42 1969 (1) pp.149–58.

'In the footsteps of Samuel Smiles' *The Author*, Journal of the Society of Authors, Winter 1970, pp.151–55.

Rolt, Tim, 'A Short History of the Rolt family at Stanley Pontlarge', unpublished essay, nd.

Uren, C., 'No Room for Complacency' *Talyllyn News*, No.2, November 1953, pp.2–4.

Willans, Kyrle, 'Peter William Willans, 1851 – 1892', *Transactions of the Newcomen Society* 28 (1951) Pt 1, pp.21–34.

'Future of an Old Canal', first published in *Country Life*, 24 December 1948; repr. *Basingstoke Canal News*, Autumn 2013, pp.18–20.

Selected Online Sources

https://archive.org
www.britishnewspaperarchive.co.uk
www.canalworld.net
www.durseyglos.org.uk
https://hansard.parliament.uk/
www.hathitrust.org
www.kerrstuart4415.org.uk
www.ltcrolt.org.uk.
www.motorsportmagazine.com/archive
www.railwaysarchive.co.uk
www.steamindex.com
www.wikipedia.org
www.talyllyn.co.uk

Index

Aickman, Ray (Edith Ray Gregorson)
   literary agent, 75, 80, 98, 100, 102-3, 126
   Robert Fordyce, writer and literary agent, 75-85, 87-94, 95, 98-107, 163-4, 166
Ailsa Craig, 89-90, 106
'Anna', *see* 'Cara'
Aldbourne Engineering Company, 43-44, 66-67, 69-71, 75, 97
Allen and Unwin Ltd. publishers, *see* Unwin, Philip
Anglo-American Car Rally (1954), 131-3
Armstrong, Major Herbert Rowse, solicitor of Hay on Wye, 15
Armstrong, Sir William, armaments manufacturer, 143
Awdry, Rev. Wilbert, 125-6, 165

Banbury, 54-5, 58-9, 61, 67, 69, 72, 78, 80-1, 87, 93, 95, 98, 105-6, 109, 112, 119
Bailey, Joe, at Pitchill, 21
Basingstoke Canal, 92-3
Batsford, Harry, publisher, 71, 146
Beech, Mr, boatbuilder, 34, 96
Bennett, I. A., of Hungerford, 43
Betjeman, Sir John, poet, 102, 127, 154, 165-6
Birks, Aubrey, VSCC member, 54

Blewitt, Joseph, trades union official, 101
Bloxham, Alan, at Pitchill, 21
Bomford of Pitchill, agricultural engineers, 20-29 *passim*
   Benjamin, 20
   Douglas, 25, 27
   Ernest, ('Hercs') 21
   Leslie, 27, 30, 39, 40
   Raymond, 20-21
Boucher, Roy, solicitor of Bristol, 142
Bowen, John, VSCC member, 54, 58, 63, 120
Bowen, Major Sir Edward Crowther, 54, 58
Boyd, James I. C., early Talyllyn volunteer, 117
Braunston, 54-5, 87, 89, 120, 122
Brindle, Steven, biographer, 144-5
Brindley, James, engineer, 104, 143, 159
Brunel, Isambard Kingdom, engineer, 2, 141-7, 152, 159, 166-7
Bryneglwys Slate Quarries, 110-12
Buchanan, Angus, early writer on Industrial Archaeology, 144-5, 154
Burnyeat Ltd., *see Mariners' Market*

California Works, *see* Kerr Stuart & Co, Ltd.
Canal Art Exhibition, LTCR organises, 87-8, 104

Canal du Midi, *see From Sea to Sea*
Cape, David, publisher, 141
  Jonathan, publisher, 141-2
'Cara' aka 'Anna', 50-53, 55
Carson, Tim, VSCC member, 46-8
  Peter, publisher, 166
Cavender, A.S., IWA member, 106
Clarke, Augusta, *see* Taylor, Augusta
Clarke, Jemima, *see* Rolt, Jemima, known as 'Mima'
Clarke, Mary, *see* Timperley
Clarke, Thomas of Cogshall Hall, 4, 5
Clutton, Cecil Samuel, VSCC member known as Sam, 47-8, 69, 165
Chambers, Aidan, writer of ghost stories, 164
Corris Railway, 114, 116, 127
Coster, Howard and Joan, photographers, 75
Cressy, 34-5, 53-6, 58-9, 61-63, 65, 67-70, 72, 75, 77-8, 80, 83-6, 89, 93, 95-100, 102-5, 107-110, 118-19, 122, 163, 166
Curwen, Barbara, née Willans, 114, 166
  David, 96-7, 110, 112, 114-5, 117, 119-20, 125, 133

David & Charles, publishers, 151, 158
Davis, Sidney, IWA solicitor, 100
Davies, Mrs, refreshment lady, 126
Delgove, Henri, aviation enthusiast, 142
Dolgoch (locomotive), *see* Talyllyn Railway – Locomotives
Dolgoch (place), 117-18, 120, 125, 127
Dominican Republic, the Rolts' visit to, 148-49

Edward Thomas, *see* Talyllyn Railway – Locomotives
Edwards, Lewis, known as 'Teddy', 92, 106
Edye, T. H., friend of LTCR's at Dursley, 42
Emily Burnyeat, ship owned by Burnyeat Ltd., 150
Euston Arch, demolition of, viii, 153
Eyre, Frank, friend of Charles Hadfield, 76, 134

Fairbourne Railway, 118
Fairgrieve, John, vascular surgeon, 160
Fortune, E. J, professional name 'John Fortune,' Leicester-based journalist, 53-4, 62, 104

Garfield, Simon, author of *The Last Journey of Willian Huskisson*, 145
Garland, Patrick, accountant, active in Talyllyn Railway Preservation Society, 114
Garnett, Annie, *see* Rolt, Annie
Garnett, Henry, LTCR's great grandfather and associate of James Nasmyth, 1
Garnett, Rev. Lionel, Rector of Christleton, 3
Garraway, Allan, railwayman and Talyllyn volunteer, 129
George Cohen & Co Ltd., manufacturers and scrap dealers, 37
Glover, Julian, biographer of Thomas Telford, 145
Grant-Ferris, Robert, Conservative politician, 160
Greenbergh, G. R., Kerr Stuarts' sales manager, 32
Gregorson, Edith Ray, *see* Aickman, Ray

Index

Greyfriar's House, Chester home of George and Augusta Taylor, 12
Griffith, Hugh, Welsh actor, married to Gunde, 96
Grundy family, early canal enthusiasts, 84-6, 95, 108
Guest, John, writer and publisher, 130, 142, 144, 224
Gwynne, Mary, LTCR's nurse, 6

Hadfield, Charles, canal historian, 76, 77, 79, 80, 84, 87, 88, 96, 134, 151, 157-58
Hadlow, Charles, canal enthusiast, 87
Hardinge, George, publisher, 142
Hardy, Major Henry, LTCR's headmaster, 18-19
Harper, Dick, GP and friend of LTCR's, 161, 165, 166
Harvey, Bill, railwayman and Talyllyn volunteer, 128-29
Herbert, A. P., *see* Herbert, Sir Alan
Herbert, Sir Alan, distinguished writer, MP, and President of the IWA, 81, 87, 104-5
Hopthrow, Brigadier Harry Ewart, 105
Hough, Richard, publisher, 134
Houldsworth McConnel family, *see* Bryneglwys Slate Quarries
Howard, Elizabeth Jane, novelist, 79, 88-9, 91, 102
Huddersfield Narrow Canal, 89-91
Hunslet Engineering Company, 38, 151
Hudson Point, bulk carrying cargo ship, 148-9
Humm, Robert, Talyllyn volunteer, 117

Inland Waterways Association, formation of, 77
Inland Waterways Redevelopment Committee, 154

Jessop, William, canal engineer, 146, 158
Jones, Dai, employed on the Talyllyn Railway, 116-7
Jones, Sir Henry Haydn, Ironmonger, MP and owner of the Talyllyn Railway, 111-14, 116, 119
Jones, Hugh, employed on the Talyllyn Railway, 116

KS 4415, early diesel locomotive, 32
Karslake, Kent, VSCC member, 48
Kennet Navigation, 70
Kennet and Avon Canal, 66, 70
Kerr Stuart & Co. Ltd, 29-39 *passim*
Kilvert, Rev. Francis, diarist, 148

Lamb, Hugh, writer of ghost stories, 164
Langham Reed, Herbert, Director of Kerr, Stuart & Co Ltd., 35-6, 38
Lardner, Dr Dionysius, 19th century writer on science and engineering, 134, 145
Lester, Percy, employed at Pitchill, 21
Leonard, George, employed at Pitchill, 21, 24
Lifford Bridge, 82, 84, 105
Lines, Ernie, employed at Kerr Stuart & Co Ltd, 30, 32
Lister, Sir Ashton, founder of R. A. Lister & Co., Dursley, 40
Listers of Dursley, *see* R. A. Lister & Co. Ltd
Livock, Squadron Leader Gerald, IWA member, 86
Lloyd, Rev. Canon Roger, railway writer and book reviewer, 127
Longman, publisher, 130, 142, 154
Longman's Industrial Archaeology Series, 154

209

Lunar Society, 155
Lycett, Forrest, VSCC member, 48

Mackersey, Ian, Rolt enthusiast and writer, vi, viii, 62, 64, 68, 78, 107
Maconie, Stuart, Rolt enthusiast and DJ, 167
Maguire, Bill ('Maggie') on the Talyllyn Railway staff, 116
Malloch, Ron, friend of LTCR's at Dursley, 42-3
Market Harborough, 98, 102-106, 113, 163
Marples, Ernest, 154
Marrian, Ken, General Manager of the Talyllyn Railway 1953-8, 128-29
Marshall, Joan, purchaser of Basingstoke Canal, 92
McMullen, Col Dennis, of the Railways Inspectorate, 134
Miller, Daphne, works for IWA, 79
Morse, Greg, railway writer, 141
Mosley, Lady Cynthia, MP for Stoke-on-Trent, 38
Mystery Stories magazine, 48, 55

National Railway Museum, 154
Newman Neame Ltd., publisher, 147, 150-51
Noble, Lady Celia Brunel née James, grand-daughter of Isambard Brunel, 142
Noble, Major Sir Humphrey, son of Lady Celia Brunel Noble and great-grandson of Isambard Brunel, 142-3
Celia, née Weigall, Sir Humphrey's wife, 142
Nurser, Frank, boat painter, 87, 89

Odiham, 50
Oliver, Bill, Talyllyn Railway Staff, 116, 126
Orred, Angela *see* Rolt, Angela
Cynthia, née Fitz-Clarence, Angela Rolt's mother, 57-8, 60, 107
Diana, Angela Rolt's sister, 57
Major George Roland, Angela Rolt's father, 54, 57-61, 107

Perry, Captain John, works at Dagenham, 147
Pitchill, 20-28, *passim*
Phoenix Garage, Hartley Wintney, 46, 52-4, 56
Plas-y-Garth, Welsh home of George and Augusta Taylor, 4, 13
Pontcysyllte Aqueduct, 34, 61, 84-5, 95-6

R. A. Lister & Co. Ltd., agricultural machinery manufacturer, 40-49 *passim*
Ravenglass and Eskdale Railway, 32, 97
Robbins, Will, at Pitchill, 21-3, 25
Rogers, Harry, coracle maker, 44
Rolls Royce, Crewe, 65-8, 73, 75, 155
Rolt, Algernon, 5, 10
Angela, née Orred, LTCR's first wife, viii, 54-65, 70-73, 77-80, 84-91, 95-6, 98-100, 102-8, 110-11, 113, 119-20, 122, 127, 162, 164, 166
Annie, née Garnett, LTCR's grandmother, 2, 3, 6
David, LTCR's cousin; landscape and portrait painter, 109, 122, 165
Dorothy, LTCR's aunt, 5
Gladys, LTCR's aunt, 5
Harry, LTCR's uncle, 5-6

Index

Hilda, LTCR's aunt, 5
Col Sir John, LTCR's great-grandfather, 1-2, 5
Rolt, L. T. C. (Lionel Thomas Caswall) known as Tom, *passim*
Published Books (This list includes only those titles which the text mentions)
A Company's Story in its Setting: Samuel Williams & Sons Ltd 1855-1995, 147-150
From Sea to Sea, 160
George and Robert Stephenson – The Railway Revolution, 143, 145-7, 152, 159, 167
Great Engineers, 146
Green and Silver, 53, 79, 80, 119
High Horse Riderless, 71, 84, 96, 102, 167
A Hunslet Hundred, 151
Isambard Kingdom Brunel, 141-7, 152, 159, 166-7
The Inland Waterways of England, 88, 96, 103, 119
James Watt, 146-7
Landscape with Machines, viii, 1, 6, 7, 16-17, 19, 29-30, 34, 36, 40-41, 46, 50, 57, 63, 161, 163
Landscape with Canals, viii, 1, 61, 63, 91-2, 102, 104, 119, 163
Landscape with Figures, viii, 111, 117-8, 122, 124, 129, 131, 148, 150, 152, 154, 161, 165
Lines of Character, 13, 123-25, 130, 137, 140-41, 167
Mariners' Market, 150
Narrow Boat, viii, 63, 68-9, 73-5, 76-8, 102, 127, 146, 152, 167
Navigable Waterways, 153-4, 159
Patrick Stirling's Locomotives, 153
Railway Adventure, 97, 111, 118-9, 120, 127, 130, 141
Red for Danger, vi, 24, 34, 79, 134-5, 138, 140-41, 152, 167
Sleep No More, 31, 80, 164, 167
The Thames from Mouth to Source, 102
Thomas Telford, 145
Turner and Newall Limited – The first fifty years 1920-1950, 152
Victorian Engineering, 153, 155
Waterloo Ironworks – A History of Taskers of Andover, 1809-1968, 151
Winterstoke, 130, 141, 167

Manuscripts
'Strange Vista' (Unpublished novel), 31, 44-6, 51, 55, 130, 155
'The Light Railway of Today' (Runner-up in Stoke-on-Trent Association of Engineers' Essay Competition, 1929), 32
'The Railway' (Collection of early poems, stories and other short pieces), 13-14, 126
'This Modern Generation – A Response to its Critics' (Essay about patriotism and pacifism), 64-5

Rolt, Jemima, known as 'Mima', née Clarke; later Clarke-Timperley, LTCR's mother, 4-5, 11, 16-17, 56-7, 68, 72, 80, 94, 100, 127, 129
Lionel -'Lily', LTCR's father, 3-12, 16-19, 56, 64, 72
Marjory, LTCR's aunt, 5
Neville, LTCR's uncle, 5, 16
Richard, LTCR's son, vi, 108, 127-9, 132, 159-60, 165

Sonia, née South, later Smith,
  LTCR's second wife, 74,
  80-2, 99, 101-2, 104, 106-8,
  119-25, 127-9, 132-3, 148-9,
  159-66
Thomas Francis, LTCR's
  grandfather, 2-3
Tim, LTCR's son, vi, 108, 127,
  152, 157, 159-60, 165
Wilfred, LTCR's uncle, 5, 16
Rose, Harry, school friend of LTCR,
  18, 71
Rudman, Frank, publisher, 141
Russell, Jim, rail-enthusiast friend of
  LTCR, 95, 96, 112, 115
Russell-Cooke, [Helen] Melville,
  'Mel', 109, 122

San Domingo, 149
Schonell, Professor Fred Joyce,
  Birmingham-based educationalist,
  99, 101
Science Museum Advisory
  Council, 154
Scott, Peter, painter, naturalist
  and canal enthusiast, 80, 102,
  104-5, 163
Scott Russell, John, 143-4
Sentinel Waggon Works Ltd.,
  Shrewsbury, 27, 44
Shropshire Union Canal (Welsh
  Section), 61, 67, 82, 84, 95, 166
Sir Haydn, see Talyllyn Railway –
  Locomotives
Sitwell, Sacheverall, 104
Skempton, Sir Alec, 158
Smeaton, John, 158
Smiles, Samuel, 44, 144, 147
Smith, Bill at Pitchill, 21, 24, 26,
  27, 44
Smith, Sir John and Christian,
  162, 165

Smith, George, boatman and Sonia
  Rolt's first husband, 74, 80, 82,
  104, 122
Sonia, *see* Rolt
Snell, John, Talyllyn volunteer, 111,
  116-7, 120, 159
South, Sonia, *see* Rolt
Southwick, Thomas, LTCR's tutor,
  6-8, 15, 17
Springtime for Henry, play by Benn
  Levy staged at Market Harborough
  Rally, 102
Standedge Tunnel, 89
Stanley Pontlarge, 11, 13, 15, 20, 43,
  48, 59, 72-3, 109, 122, 127, 129,
  133, 143, 159, 165-6
Stephenson, George, 146-7
Stratford-upon-Avon Canal, 75,
  81-2, 95
Sutherland, James and Anthea, 89-91

Talyllyn, see Talyllyn Railway –
  Locomotives
Talyllyn Railway, 72, 96, 109, 110-21
  passim, 123, 127-8, 151, 166
Talyllyn Railway – Locomotives
  No. 1 Talyllyn, 97, 110, 112,
    115, 118
  No. 2 Dolgoch, 97, 110-11, 115,
    118-9, 129
  No. 3 Sir Haydn, 116, 128-9
  No. 4 Edward Thomas, 116, 129
Talyllyn Railway Preservation
  Society, 114-7, 119, 121, 125, 128
Tardebigge, 72, 74-5
Taylor, Augusta, née Clarke, later
  Clarke-Timperley, LTCR's aunt,
  4, 7-8
George, 4, 7-8
Hero, *see* Willans
Temple Thurston, E., 151
Telford, Thomas, 96, 104, 145-7, 151

212

Index

Timperley, Herbert Luard, 8
Timperley, Rev Isaac Robinson, 4-5, 11
Timperley, Mary, previously Clarke, LTCR's grandmother, known as 'Granny Timperley', 9
Thomas, David St John, 151, 158
Thomas, Edward, 112-16, 125
Thomas, Jimmy, MP, 36
Thornycroft of Basingstoke, 44, 46, 65, 93
Tooley, family of boatbuilders and boat painters, 58-9, 61, 78, 80, 87
Tooley's Yard, Banbury, 67-8, 70, 105
Tower, Geoffrey and Maria, 98
Towyn, 72, 97, 109-19, 125-6, 128
Tralee and Dingle Railway, 123-24
Transport and General Workers' Union, see Blewitt, Joseph
Tredgold, Thomas, 154
Trent and Mersey Canal, 29, 34, 62-3, 67, 89, 92
Trevithick, Richard, 146, 167
Trinder, W. H., 'Bill', 95-7, 105, 109, 112-14, 119
Tywyn, see Towyn

Unwin, Philip, publisher, 70-71, 88, 103, 110, 116, 119, 123, 125
Uren, Charles, engineer associated with Talyllyn Railway, 129

Van der Gucht, Mrs, see Vanderguilt, Mrs
Vanderguilt, Mrs, housekeeper at The Towers, Dursley, 42
Vaughan, Adrian, 144-5
Vintage Sports Car Club, see VSCC
VSCC, 46-8, 53-6, 59, 69, 87, 120, 122, 131-32, 144, 161, 165, 167

Waldman, Milton, biographer and publisher, 142
Walwick Hall, home of Sir Humphrey and Lady Noble, 142-3
Ward Lock, publishers, 158-59
Watt, James, 146-47
'Welsh Canal', see Shropshire Union Canal (Welsh Section)
Williams, Samuel, 147-48
Whitehouse, Patrick, active in Talyllyn Railway Preservation Society, 114, 123, 129
Wilkins, John, active in Talyllyn Railway Preservation Society, 118
Willans, Barbara, also known as Barbara Curwen, 114, 166
Bill, LTCR's cousin, 33-4, 54, 56, 90
Hero (née Taylor) LTCR's godmother, 4, 20, 84-5, 87, 94
Kyrle, LTCR's uncle, 4, 20, 23, 27, 30, 32, 33-9, 53-4, 56, 84-5, 93-4, 96, 100, 108, 162
Wyatt, Rendel, founder of Canal Cruising Company, 89, 106, 108